Antibacterial Treatment in Periodontal and Endodontic Therapy, 2nd Edition

Antibacterial Treatment in Periodontal and Endodontic Therapy, 2nd Edition

Guest Editors

Andreas Braun
Felix Krause

MDPI

Basel • Beijing • Wuhan • Barcelona • Belgrade • Novi Sad • Cluj • Manchester

Guest Editors
Andreas Braun
Clinic for Operative Dentistry,
Periodontology and
Preventive Dentistry
RWTH Aachen University
Aachen
Germany

Felix Krause
Clinic for Operative Dentistry,
Periodontology and
Preventive Dentistry
RWTH Aachen University
Aachen
Germany

Editorial Office
MDPI AG
Grosspeteranlage 5
4052 Basel, Switzerland

This is a reprint of the Special Issue, published open access by the journal *Antibiotics* (ISSN 2079-6382), freely accessible at: https://www.mdpi.com/journal/antibiotics/special_issues/882LQ5C8MX.

For citation purposes, cite each article independently as indicated on the article page online and as indicated below:

Lastname, A.A.; Lastname, B.B. Article Title. *Journal Name* **Year**, *Volume Number*, Page Range.

ISBN 978-3-7258-5985-6 (Hbk)
ISBN 978-3-7258-5986-3 (PDF)
https://doi.org/10.3390/books978-3-7258-5986-3

Contents

About the Editors vii

Preface ix

Patrick Jansen, Georg Conrads, Johannes-Simon Wenzler, Felix Krause and Andreas Braun
Bacteremia Prevention during Periodontal Treatment—An In Vivo Feasibility Study
Reprinted from: *Antibiotics* **2023**, *12*, 1555, https://doi.org/10.3390/antibiotics12101555 1

Ewa Dolińska, Anna Skurska, Violetta Dymicka-Piekarska, Robert Milewski and Małgorzata Pietruska
Matrix Metalloproteinase 9 (MMP-9) and Interleukin-8 (IL-8) in Gingival Crevicular Fluid after Minimally Invasive Periodontal Surgery with or without Er:YAG and Nd:YAG Laser Application
Reprinted from: *Antibiotics* **2024**, *13*, 704, https://doi.org/10.3390/antibiotics13080704 13

Sarah Böcher, Hendrik L. Meyer, Evdokia Dafni and Georg Conrads
Prevalence and Phylogenetic Analysis of Lipoprotein-Gene *ragB-1* of *Porphyromonas gingivalis*—A Pilot Study
Reprinted from: *Antibiotics* **2023**, *12*, 1458, https://doi.org/10.3390/antibiotics12091458 27

Yasuo Takeuchi, Akira Aoki, Koichi Hiratsuka, Chanthoeun Chui, Akiko Ichinose, Nay Aung, et al.
Application of Different Wavelengths of LED Lights in Antimicrobial Photodynamic Therapy for the Treatment of Periodontal Disease
Reprinted from: *Antibiotics* **2023**, *12*, 1676, https://doi.org/10.3390/antibiotics12121676 39

Johannes-Simon Wenzler, Wolfgang Falk, Roland Frankenberger and Andreas Braun
Temporary Root Canal Obturation with a Calcium Hydroxide-Based Dressing: A Randomized Controlled Clinical Trial
Reprinted from: *Antibiotics* **2023**, *12*, 1663, https://doi.org/10.3390/antibiotics12121663 59

Dirk-Joachim Drews, Anh Duc Nguyen, Antje Diederich and Christian Ralf Gernhardt
The Interaction of Two Widely Used Endodontic Irrigants, Chlorhexidine and Sodium Hypochlorite, and Its Impact on the Disinfection Protocol during Root Canal Treatment
Reprinted from: *Antibiotics* **2023**, *12*, 589, https://doi.org/10.3390/antibiotics12030589 70

Lauzan haj Khalaf, Salma Kabbaj and Babacar Toure
Attitudes of Dental Students towards the Prescription of Antibiotics during Endodontic Treatment
Reprinted from: *Antibiotics* **2024**, *13*, 913, https://doi.org/10.3390/antibiotics13100913 88

Koay Chun Giok and Rohit Kunnath Menon
The Microbiome of Peri-Implantitis: A Systematic Review of Next-Generation Sequencing Studies
Reprinted from: *Antibiotics* **2023**, *12*, 1610, https://doi.org/10.3390/antibiotics12111610 96

About the Editors

Andreas Braun

Andreas Braun is affiliated with the Clinic for Operative Dentistry, Periodontology, and Preventive Dentistry at RWTH Aachen University Hospital, where he has served as Head and Chair since 2019. He completed his dental studies at the University of Bonn, where he also earned his doctorate and began his academic career. After holding teaching and research positions at the Universities of Bonn and Marburg, he became Chief Consultant of the Department of Operative Dentistry and Endodontology at Philipps University of Marburg from 2011 to 2019. His professional interests include periodontology, endodontology, restorative and preventive dentistry, and the clinical application of laser technologies. Professor Braun has been Head of the TransMIT project division for energy-transmitting applications in dentistry since 2015 and has contributed significantly to advancing evidence-based and technology-enhanced dental care. Internationally recognized for his scientific and clinical work, he currently serves as President of the German Society for Laser Dentistry and, since 2021, as Head of the Master's program "M.Sc. Periodontology" at RWTH Aachen University.

Felix Krause

Felix Krause is currently Chief Senior Physician and Deputy Clinic Director at the Department for Operative Dentistry, Periodontology, and Preventive Dentistry at University Hospital RWTH Aachen. He studied Dentistry at Freiburg University from 1992 to 1999 and earned his doctoral degree in 2002. Between 1999 and 2009, he served as a Research Associate at the Department of Operative Dentistry and Periodontology and later at the Department of Periodontology, Operative, and Preventive Dentistry at the University of Bonn, where he completed his habilitation in 2007 on laser fluorescence applications in dental diagnostics. From 2009 to 2011, he led the Department of Knowledge Communication at VOCO GmbH in Cuxhaven, and subsequently served as Chief Senior Physician, Deputy Clinic Director, and Head of the Division of Preventive Dentistry at Leipzig University Hospital from 2011 to 2019. He completed postgraduate studies in Medical Education at Heidelberg University (MME, 2017) and received the Dental Education Award of the Kurt-Kaltenbach Foundation. His professional interests focus on preventive dentistry, laser-based diagnostics, and innovations in dental education. From 2019 to 2023, he was a board member of the Working Group on the Advancement of Dental Education in Germany (AKWLZ) and, since 2021, he has been a lecturer in the International Postgraduate Master's Program "Periodontology" of RWTH Aachen. In 2022, he was appointed Extraordinary Professor at the Faculty of Medicine, RWTH Aachen.

Preface

In recent years, the fields of periodontal and endodontic therapy have witnessed significant advancements, particularly concerning antibacterial treatment protocols. As we delve into this special edition of the Journal *Antibiotics*, we aim to provide a comprehensive overview that not only highlights current research but also addresses contemporary questions that are pivotal to both scientific inquiry and clinical practice.

The purpose of this special edition is twofold: first, to present new and intriguing information on antibacterial treatments as they relate to periodontal and endodontic therapies; second, to foster a dialogue around pressing issues that practitioners face in real-world settings. With antibiotic resistance becoming an increasingly urgent global health concern, it is critical that we explore innovative approaches and alternative therapies that can enhance patient outcomes while mitigating potential risks associated with traditional antibiotic use. In today's fast-moving world, knowledge within the scientific community is evolving at an unprecedented pace. This rapid advancement necessitates a continuous exchange of ideas between researchers and clinicians. By bridging the gap between laboratory findings and clinical applications, this special edition reflects our commitment to ensuring that practitioners have access to up-to-date information that can directly influence their treatment strategies.

The articles featured in this issue encompass a diverse range of topics, from novel antibacterial agents to emerging techniques in microbial management. We encourage both researchers who are pushing the boundaries of knowledge and clinicians who are applying these insights in practice to engage with the content presented here. Our hope is that this special edition will serve as a valuable resource for professionals seeking evidence-based guidance amid evolving challenges in oral health care. As we navigate through complex cases involving periodontal disease and endodontic infections, it is imperative that we remain informed about current trends, innovative treatments, and best practices. The discussions initiated by this collection of articles will undoubtedly contribute to shaping future research directions while enhancing clinical decision-making processes.

We invite you to immerse yourself in this special edition and participate in ongoing conversations about how we can collectively improve antibacterial treatment strategies within periodontal and endodontic therapy. Together, let us advance our understanding and application of these crucial interventions for better patient care outcomes. We hope that we have selected an interesting compilation of scientific papers for you, with which we aim to achieve excellence in dentistry through cooperation between research and practice.

Andreas Braun and Felix Krause
Guest Editors

antibiotics MDPI

Article

Bacteremia Prevention during Periodontal Treatment—An In Vivo Feasibility Study

Patrick Jansen [1,*], Georg Conrads [2], Johannes-Simon Wenzler [1], Felix Krause [1] and Andreas Braun [1]

[1] Clinic for Operative Dentistry, Periodontology and Preventive Dentistry, Rheinisch-Westfälische Technische Hochschule University Hospital, Pauwelsstrasse 30, 52074 Aachen, Germany; jwenzler@ukaachen.de (J.-S.W.); fkrause@ukaachen.de (F.K.); anbraun@ukaachen.de (A.B.)

[2] Division of Oral Microbiology and Immunology, Clinic for Operative Dentistry, Periodontology and Preventive Dentistry, Rheinisch-Westfälische Technische Hochschule University Hospital, 52074 Aachen, Germany; gconrads@ukaachen.de

* Correspondence: pjansen@ukaachen.de; Tel.: +49-241-80-88110

Abstract: The link between periodontitis and systemic diseases has increasingly become a focus of research in recent years. In this context, it is reasonable—especially in vulnerable patient groups—to minimize bacteremia during periodontal treatment. The aim of the present in vivo feasibility study was to investigate the possibility of laser-based bacteremia prevention. Patients with stage III, grade B generalized periodontitis were therefore treated in a split-mouth design either with prior 445 nm laser irradiation before nonsurgical periodontal therapy or without. During the treatments, clinical (periodontal measures, pain sensation, and body temperature), microbiological (sulcus samples and blood cultures before, 25 min after the start, and 10 min after the end of treatment), and immunological parameters (CRP, IL-6, and TNF-α) were obtained. It was shown that periodontal treatment-related bacteremia was detectable in both patients with the study design used. The species isolated were *Schaalia georgiae*, *Granulicatella adiacens*, and *Parvimonas micra*. The immunological parameters increased only slightly and occasionally. In the laser-assisted treatments, all blood cultures remained negative, demonstrating treatment-related bacteremia prevention. Within the limitations of this feasibility study, it can be concluded that prior laser disinfection can reduce bacteremia risk during periodontal therapy. Follow-up studies with larger patient numbers are needed to further investigate this effect, using the study design presented here.

Keywords: periodontal treatment; periodontal therapy; nonsurgical periodontal therapy; bacteremia; laser; laser disinfection

1. Introduction

Periodontitis is an inflammatory disease of all structures of the periodontium, maintained by microorganisms in the subgingival biofilm. Inflammatory changes can lead to loss of bone, collagen, and attachment. In advanced disease and without adequate treatment, the consequence may be loss of the affected teeth. Irreversible periodontal damage is usually preceded by reversible gingivitis due to dysbiotic microbial colonization of the tooth and/or gum.

In addition to this basic condition, systemic diseases represent cofactors for periodontitis, and some of these may in turn be influenced by the periodontitis itself. These systemic diseases include, for example, diabetes mellitus, cardiovascular disease (such as coronary artery disease), and prematurity [1–3].

In particular, the bidirectionality between diabetes mellitus and periodontitis has been well studied. Various clinical studies have shown that periodontitis is more common in diabetics than in non-diabetics [4–6]. Periodontitis, in turn, may also influence diabetes mellitus. For example, the risk of diabetics dying from coronary heart disease or nephropathy is significantly higher if they have advanced periodontitis [7].

In addition to diabetes, many studies have also shown an association between periodontitis and coronary heart disease [2,8,9].

Recently, in addition to the systemic diseases already mentioned, a link between periodontitis and Alzheimer's disease (AD) has also been discussed [10–13]. A prospective clinical study showed that active, chronic periodontitis was associated with a decline in mental abilities in nursing home residents [14]. In particular, *Porphyromonas gingivalis* (*P. gingivalis*), one of the main pathobionts involved in periodontitis, appears to play a key role in this regard [11,15,16]. Thus, it is clear from these examples that periodontal health is closely related to systemic health.

Removal of the soft biofilm or mineralized deposits adhering to the tooth surface and thus the removal of periodontal pathogenic microorganisms (pathobionts) by scaling and root planing (SRP) is one of the main objectives of nonsurgical periodontal therapy and supportive periodontal therapy [17], regardless of the systemic factors that influence the event. Conventional methods such as the use of manual or ultrasound instruments are available for this purpose, and their effectiveness is considered to be comparable [18,19]. Another option is the use of air polishing systems or laser systems for cleaning affected tooth surfaces [20–23]. However, the benefit of root surface cleaning in periodontal therapy, e.g., with the Er:YAG laser, remains questionable due to the high heterogeneity of clinical studies [24]. Generally, it is possible to achieve a reduction in periodontal pathobionts using the various methods of surface instrumentation [25,26]. It cannot be assumed that complete removal of pathogens is achieved. Even the additional local application of an antibiotic—for example, using tetracycline-releasing sutures [27] or doxycycline-releasing gels [28,29]—cannot guarantee that all pathogens will be permanently removed. Furthermore, especially in advanced cases with high probing depths ($\geq$6 mm), furcation involvement or, for example, root concavities, it becomes difficult or even impossible to perform this cleaning successfully through nonsurgical approaches alone; here, it then makes sense to consider surgical approaches (resective or regenerative) [30].

During such cleaning of the tooth or root surfaces, which is an invasive procedure, there is a risk of bacteremia [31–33]. In most cases, this bacteremia has no (or unnoticed) clinical relevance in the majority of patients. However, such bacteremia can cause dangerous infections in patients who have heart valve replacements or are immunosuppressed. Here, antibiotic prophylaxis and/or removal of the pathobionts makes all the more sense [34]. A direct effect of these bacteremia during nonsurgical periodontal therapy on other systemic diseases, as mentioned above, has not yet been proven. However, this topic itself should continue to be critically observed scientifically.

Another effective approach to the elimination of pathobionts could be the use of light energy, in the form of lasers. Diode lasers achieve particularly good results in adjunctive periodontal therapy [35,36]. They also appear to lead to the effective elimination of periodontal microorganisms. One study showed, in a split-mouth design, that the use of a diode laser (810 nm) before ultrasound scaling in patients with gingivitis resulted in a significant reduction in the prevalence of odontogenic bacteremia [37].

The severity of periodontal disease can also be exacerbated by specific genotypes that influence the immune response, with a well-known example being the interleukin-1 polymorphism. Likewise, periodontitis may start significantly earlier in such patients than in patients without this polymorphism [38]. In addition to the bactericidal effects of lasers, it is also plausible to consider photobiomodulatory effects, especially with blue and green laser light, which may have a positive influence on the immune response and post operative inflammation and wound healing [39]. A current approach in this context involves the use of a novel diode laser with a wavelength of 445 nm. In an in vitro study, this wavelength was shown to effectively reduce a broad spectrum of pathobionts, including *Aggregatibacter actinomycetemcomitans*, *P. gingivalis*, *Staphylococcus aureus*, and *Candida albicans*, with a relatively short application time and low total energy [40]. This approach needs to be investigated further. The aim of the present feasibility study was therefore to evaluate the bacteremia-preventive effect of disinfection with a diode laser before conventional nonsurgical periodontal therapy—testing the hypothesis that a bacteremia-preventive effect of laser irradiation prior to periodontal treatment can be demonstrated with the study design used.

2. Results

The present feasibility study demonstrated that the study design is capable of detecting pathobionts that had entered the bloodstream due to therapy-induced bacteremia. The laser parameters used also demonstrated a bacteremia-preventive effect.

2.1. Clinical Parameters

2.1.1. Patient 1

The patient had stage III, grade B generalized periodontitis. Probing pocket depths of up to 6 mm were present, the highest degree of mobility was I (teeth 25 and 45), and the highest degree of furcation involvement was also I (teeth 17, 16, 26, 37, and 36). Periodontal recessions were mainly present in the mandibular anterior region, but these were only slight (recession type I: maximum 1 mm).

At the first appointment, the patient had a score of 0 on the visual analog scale (scores of 0–10) for laser treatment (inactive laser) and a score of 6 for subsequent SRP. At the second appointment, the scores were 3 on the visual analog scale for laser treatment (active laser) and 4 for subsequent SRP.

Measurement of the patient's body temperature (in the ear) showed the following values at the first appointment: before treatment: 36.6 °C; after treatment: 36.6 °C. Measurement of the patient's body temperature showed the following values at the second appointment: before treatment: 35.9 °C; after treatment: 36.5 °C.

2.1.2. Patient 2

The patient had stage III, grade B generalized periodontitis. Probing pocket depths of up to 10 mm (at one tooth) were present, the highest degree of mobility was I (teeth 17, 15, 14, 12–22, 25, 27, 37, 35, 31, and 41), and the highest degree of furcation involvement was II (teeth 17, 16, 27, and 46; degree I: 14, 24, 37, 36, and 47). Periodontal recessions were mainly present in the mandibular anterior region, but these were only slight (recession type I: maximum 3 mm).

At the first appointment, the patient had a score of 0 on the visual analog scale (scores of 0–10) for laser treatment (inactive laser) and a score of 2 for subsequent SRP. At the second appointment, the scores were 2 on the visual analog scale for laser treatment (avtive laser) and 2 for subsequent SRP.

Measurement of the patient's body temperature (in the ear) showed the following values at the first appointment: before treatment: 36.4 °C; after treatment: 36.1 °C. Measurement of the patient's body temperature showed the following values at the second appointment: before treatment: 36.1 °C; after treatment: 35.6 °C.

2.2. Microbiological and Immunological Parameters

2.2.1. Patient 1

Sulcus sampling identified various bacteria at the baseline examination, including *Schaalia* (formerly *Actinomyces*) *georgiae*, which was also detected in the anaerobic blood culture sample during SRP at the first appointment (see below). The complete list at each examination time point is shown in Table 1.

At the first appointment, the results of the aerobic and anaerobic blood culture analyses were as follows. The baseline examination did not reveal any anomalies or detectable transient bacteremia. In the samples taken 25 min after the start of SRP, *Granulicatella adiacens* was detected in the aerobic blood culture after several days of incubation and *S. georgiae* in the anaerobic blood culture. Samples taken 10 min after the completion of SRP did not show any anomalies, as in the baseline examination. Aerobic and anerobic blood culture analyses at the second appointment were unremarkable at the baseline examination, as well as at examinations 25 min after the start of SRP and 10 min after completion of SRP. Thus, no transient bacteremia was detected at this appointment.

Table 1. Results of the blood culture sampling and microbiological sulcus samples.

Patient 1	Time Point	Blood Culture	Assessment	Sulcus Samples (Baseline)
First treatment (inactive laser)	Baseline	Aerobic Anaerobic	Negative Negative	Dominant cultivable species: *Actinomyces israelii* *Actinomyces oris* *Capnocytophaga gingivalis* *Fusobacterium canifelinum* *Parvimonas micra* *Prevotella denticola* *Prevotella nigrescens* ***Schaalia georgiae*** *Streptococcus intermedius* *Streptococcus oralis* *Veillonella parvula*
	25 min after the start of SRP	Aerobic Anaerobic	*Granulicatella adiacens* ***Schaalia georgiae***	
	10 min after SRP	Aerobic Anaerobic	Negative Negative	
Second treatment (active laser)	Baseline	Aerobic Anaerobic	Negative Negative	
	25 min after the start of SRP	Aerobic Anaerobic	Negative Negative	
	10 min after SRP	Aerobic Anaerobic	Negative Negative	
Patient 2	**Time point**	**Blood culture**	**Assessment**	**Sulcus samples (baseline)**
First treatment (inactive laser)	Baseline	Aerobic Anaerobic	Negative Negative	Dominant cultivable species: *Actinomyces denticolens* *Actinomyces oris* *Actinomyces meyeri* *Eikenella corrodens* *Eubacterium brachy* *Fusobacterium naviforme* *Fusobacterium nucleatum* *Gemella morbillorum* ***Parvimonas micra*** *Porphyromonas gingivalis* *Prevotella denticola* *Prevotella intermedia* *Slackia exigua* *Streptococcus anginosus* *Streptococcus constellatus* *Streptococcus oralis* *Streptococcus sanguinis*
	25 min after the start of SRP	Aerobic Anaerobic	Negative ***Parvimonas micra***	
	10 min after SRP	Aerobic Anaerobic	Negative Negative	
Second treatment (active laser)	Baseline	Aerobic Anaerobic	Negative Negative	
	25 min after the start of SRP	Aerobic Anaerobic	Negative Negative	
	10 min after SRP	Aerobic	Negative	
		Anaerobic	Negative	

SRP, scaling and root planing. The bacteria names in bold refer to the bacteria that were detected both in the sulcus and later in the blood samples.

Immunological parameters—C-reactive protein (CRP) reference range < 5.0 mg/L, interleukin-6 (IL-6) reference range < 7.0 pg/mL, and tumor necrosis factor alpha (TNF-α) reference range < 8.1 pg/mL—showed the following courses in patient 1 at the first appointment: there were no changes at baseline, 25 min after the start of SRP, or 10 min after the end. One day after the first SRP, there was only a change in IL-6 (2.6 pg/mL, compared to <1.5 pg/mL at the baseline examination). The immunological parameters showed the following courses at the second appointment: at the baseline examination, the value for IL-6 was 2.96 pg/mL; at the subsequent examinations, this value decreased again to <1.5 pg/mL. The values for TNF-α were <4.0 pg/mL at the baseline examination, 4.7 pg/mL at the examination 25 min after the start of SRP, and <4.0 pg/mL again at the subsequent examinations. The values for CRP did not show any changes at the different time points. The kinetics of the inflammatory parameters are visually presented in Figure 1.

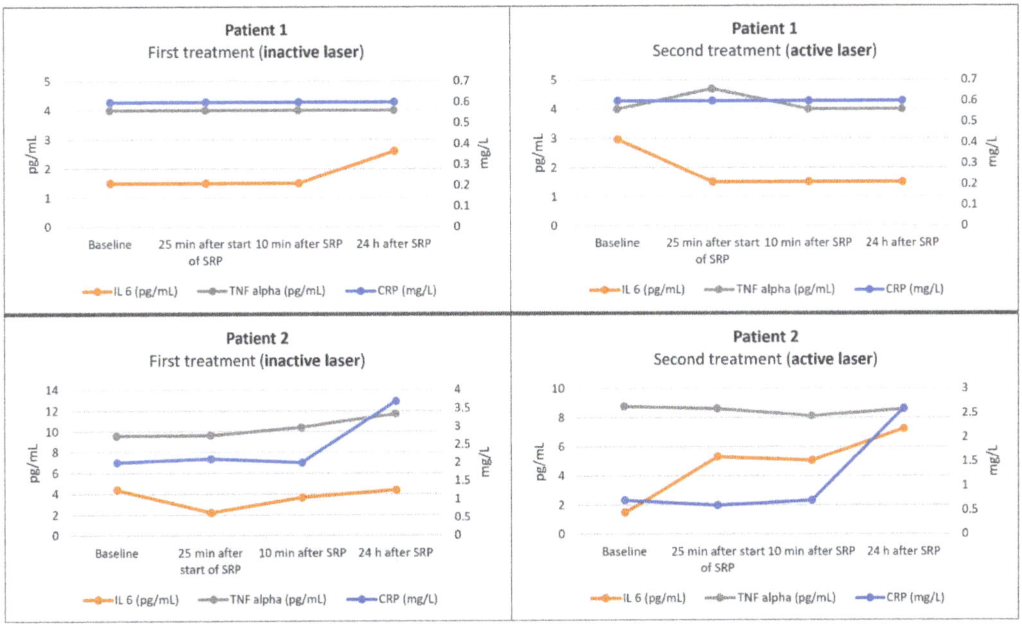

Figure 1. Diagrams of the kinetics of the inflammatory parameters (detection thresholds: IL 6: 1.5 pg/mL; TNF alpha: 4.0 pg/mL; CRP: 0.6 mg/L).

2.2.2. Patient 2

Sulcus sampling identified various bacteria at the baseline examination, including *Parvimonas micra*, which was also detected in the anaerobic blood culture sample during SRP at the first appointment (see above). *P. gingivalis* was detected in the sulcus sample at baseline, but not in the blood cultures. The complete list at each examination time point is shown in Table 1.

At the first appointment, the results of the aerobic and anaerobic blood culture analyses were as follows. The baseline examination did not reveal any anomalies or detectable transient bacteremia. In the samples taken 25 min after the start of SRP, *P. micra* was detected in the anaerobic blood culture after several days of incubation. Samples taken 10 min after the completion of SRP did not show any anomalies, as in the baseline examination. Aerobic and anerobic blood culture analyses at the second appointment were unremarkable at the baseline examination, as well as at examinations 25 min after the start of SRP and 10 min after the completion of SRP. Thus, no transient bacteremia was detected at this appointment.

Immunological parameters (CRP, IL-6, TNF-α) showed the following courses in patient 2 at the first appointment: CRP showed a slight increase 24 h after treatment (baseline: 2 mg/L; 24 h after treatment: 3.7 mg/L), but IL-6 remained more or less stable. TNF-α was already elevated above the reference range before the start of treatment, this did not change during and after treatment. At the second appointment, CRP showed a similar course (baseline: 0.7 mg/L; 24 h after treatment: 2.6 mg/L). IL-6 showed a slight increase above the reference range 24 h after treatment (7.25 pg/mL). TNF-α was elevated above the reference range before treatment, as at the first appointment, and again, this did not change during and after treatment. The kinetics of the inflammatory parameters are visually presented in Figure 1.

3. Discussion

The present feasibility study demonstrated that the use of a diode laser with a wavelength of 445 nm has potential bacteremia-preventive effects when applied immediately before SRP. However, it must be clearly stated that this effect needs to be further investigated and proven in follow-up studies with larger numbers of patients. During periodontal treatment, transient bacteremia inevitably occurs. The incidence of bacteremia depends on the degree of periodontal inflammation [31]. Kinane et al. detected bacteremia in 13% of patients after ultrasound scaling and in 20% after periodontal probing [33]. In comparison with the present feasibility study, the blood samples in that study were only taken after the periodontal procedures, so not all bacteremia may have been detected due to phagocytosis of the bacteria by immune cells. In another study, bacteremia was detected in 73.8% of patients immediately after SRP, and 19% of the blood cultures were still positive 30 min after the completion of SRP [41]. Zhang et al. detected bacteremia in 33.3% of patients 5 min after the start of SRP [42]. These studies all detected bacteremia, but with marked differences in prevalence. This may be mainly due to the different methods used and different collection times [32]. To avoid missing bacteremia as much as possible, the collection times for blood cultures in the present study design were adjusted to follow the schedule described by Beutler et al. [43]. In contrast with most studies, the present feasibility study not only investigated the presence of bacteria in blood, but also took sulcus samples to demonstrate that the blood isolates actually originated from the periodontal pockets.

It has been shown that cleaning all quadrants of a dentition within 24 h resulted in an increased acute-phase response, in comparison to cleaning quadrants in individual sessions with a time interval of 1 week between each session [44]. However, periodontal microorganisms can enter the bloodstream not only during periodontal therapy, but also as a result of professional mechanical plaque removal, or through micro-injuries from mere chewing or tooth-brushing, in accordance with the rule that the more severe the periodontitis, the more severe the bacteremia [31,43,45]. In most cases, this is assumed to have no clinical consequences (or only unrecognized ones), but in vulnerable patient groups such as patients with heart valve replacements, there may be serious sequelae. One potential sequela is infective endocarditis, which should be avoided in such cases using antibiotic prophylaxis [46]. Clear disadvantages of systemic antibiotic prophylaxis are its effects on the whole body, including the commensal microbiome and increasing resistance rates. Employing a laser device, we do not assume that complete sterilization of the periodontal pocket is possible, but we hypothesize that at least such a temporary bacterial reduction in pockets is possible with laser disinfection through which bacteremia can be avoided.

In addition to the administration of systemic antibiotics to reduce or prevent transient bacteremia, local disinfection methods may also be an option. These include, for example, mouth rinses. However, a single mouth rinse containing chlorhexidine before scaling did not show any effect on bacteremia reduction in one study [47]. In addition to oral rinses, local laser administration may also be considered. In a clinical study, a diode laser with a wavelength of 810 nm was used before ultrasonic scaling in 22 patients with gingivitis, also with a split-mouth research design. Bacteremia was detected in 15 patients after ultrasound scaling alone, and after laser treatment in only eight patients (810 nm; flexible fiber optic 300 μm; repeated beam: 0.2 s on and 0.3 s off; output power 1.0 W; laser application duration for each tooth: 15 s) before ultrasound scaling [37].

In the present study, a laser wavelength that is effective against a broad spectrum of pathobionts—including *A. actinomycetemcomitans*, *P. gingivalis*, *S. aureus*, und *C. albicans*—was used for a bacteremia reduction approach. The wavelength used was 445 nm at a power of 0.5 W, and the laser was used in continuous-wave mode with a 320 μm fiber for 1 min for each tooth. In a pilot study, these parameters showed significant reductions in *P. gingivalis* in comparison with a control group in vitro [40]. This wavelength has a very high absorption level in the spectrum of hemoglobin and pigmented bacteria, which may explain the strong reduction in *P. gingivalis*. As this was a feasibility study and data were generated from only two patients, *P. gingivalis* was detected in sulcus samples in only one patient. The bacteremia-preventive properties of the laser treatment studied here might be of interest not only for vulnerable patient groups, but also for healthy patients. As mentioned above, periodontitis and Alzheimer's disease have recently been associated. In particular, lipopolysaccharides of *P. gingivalis* have been detected in the brains of deceased patients with Alzheimer's disease [48]. The laser treatment described here might be suitable for avoiding or at least reducing this type of bacterial migration as a result of periodontal therapy, thus potentially reducing the risk of systemic diseases.

As mentioned above, Graziani et al. demonstrated that the acute-phase response was greater in full-mouth nonsurgical therapy than in quadrant nonsurgical therapy [44]. Whether the same difference existed between the two experimental groups represented in the present study was also investigated. Unfortunately, due to the small number of patients, no clear trends can be demonstrated. In patient 1, IL-6 increased slightly 24 h after conventional treatment, but this was not detected 24 h after the appointment involving laser treatment; however, there was a transient increase in TNF-α 25 min after the start of SRP, which decreased again in the subsequent examinations. In patient 2, their TNF-α levels were already elevated at baseline and did not change during the course of the examinations on both dates. The other parameters also did not show any clear trends. Follow-up studies with larger numbers of patients may clarify whether laser disinfection with the corresponding reduction in, or prevention of, bacteremia is also associated with a lower acute-phase response.

Overall, despite the small number of patients and the limitations of this feasibility study, it can be concluded that the study protocol is capable of detecting bacteremia after periodontal therapy and that the laser disinfection method described could have potential bacteremia-reducing or even bacteremia-preventive properties. This needs to be confirmed in follow-up studies, which are currently underway.

4. Materials and Methods

4.1. Clinical Parameters

As a preparatory measure, situation models of both jaws in each patient were made in order to produce individual splints. These covered the occlusal surfaces and extended to the equator of all teeth. At six locations (mesio-vestibular, central vestibular, disto-vestibular, mesio-oral, central oral, and disto-oral), milled guiding grooves were used for defined insertion of a pressure-calibrated probe (Aesculap DB764R, Aesculap, Tuttlingen, Germany) into the gingival pocket. Using the individual splints produced in this way, the pocket depths, periodontal recession, clinical attachment level, and bleeding on probing were determined at the various examination time points by one investigator (P.J.). Bleeding points occurring within 30 s of probing were documented and reported as a percentage for the entire dentition. In addition, furcation involvement and the degree of looseness were recorded.

Pain sensation during treatment was assessed by the patients using a visual analog scale after the treatment session.

Body temperature was measured before and after treatment using an infrared thermometer in the ear (Thermoscan PRO 6000, Braun, Kronberg im Taunus, Germany).

4.2. Microbiological and Immunological Parameters

For local microbiological diagnosis, samples were taken using sterile paper points (ISO 45) in the two deepest pockets of each quadrant (eight in total) at baseline, 7 days after the first appointment/SRP, at the second appointment/SRP (14 days after the first SRP), and 7 days after the second appointment/SRP. These were then analyzed by culturing on various media (with and without the addition of sheep blood) at 37 °C in appropriate atmospheric conditions and identified by Maldi-TOF (mass spectrometry).

To be able to detect systemic bacteremia after the treatments and compare it with the baseline findings, peripheral blood (25 mL) was drawn before the start of treatment, 25 min after the start, and 10 min after the end of the treatment in both the control group and the experimental group by one investigator (J.-S.W.). The patients fasted for this purpose to ensure standardization and the better accessibility of microbiological and immunological parameters. Peripheral blood cultures were analyzed using standard operational procedures by the laboratory diagnostic center at the University Hospital in Aachen, Germany. However, the blood cultures (aerobic and anaerobic) were kept for longer (maximum 21 days) and were plated several times to avoid overlooking any slow-growing species, with special emphasis on *P. gingivalis* and other black-pigmented bacteria. It should be noted that not every bacterial cell entering the bloodstream can be detected using this (or any) method, as most are very rapidly phagocytosed and thus killed and eliminated.

In addition, the samples were analyzed for immunological parameters in blood plasma to allow conclusions to be drawn concerning the acute-phase reaction. C-reactive protein (CRP), interleukin-6 (IL-6), and tumor necrosis factor alpha (TNF-α) were of particular interest. These parameters were also monitored 24 h after the respective treatment appointments.

The blood samples mentioned above were obtained from the median cubital veins. The areas of the puncture were carefully disinfected beforehand using a skin antiseptic (Octeniderm, Schülke, Norderstedt, Germany). For this purpose, the investigator responsible for blood sampling wore sterile gloves, which he kept on for the entire blood-sampling procedure. The puncture procedure was repeated at each blood collection time point. These precautions were implemented to minimize the possibility of contamination of the blood samples with skin commensals. The blood collection procedures themselves were performed with a Safety Multifly cannula (Sarstedt, Nürnbrecht, Germany). A blood culture adapter (Sarstedt, Nürnbrecht, Germany) was used for blood cultures; the aerobic blood culture bottle (Bactec Plus Aerobic/F Culture Vials 30 mL, Becton Dickinson, Franklin Lakes, NJ, USA) was inoculated first, followed by the anaerobic blood culture bottle (Bactec Lytic/10 Anaerobic/F Culture Vials 40 mL, Becton Dickinson, Franklin Lakes, NJ, USA). A blood collection system was then used for the immunological parameters (S-Monovette, Sarstedt, Nürnbrecht, Germany). The blood culture and immunological analyses were conducted in the laboratory diagnostic center at RWTH Aachen University Hospital, and the other microbiological analyses were conducted in the Division of Oral Microbiology and Immunology, Clinic for Operative Dentistry, RWTH Aachen University.

4.3. Treatment Procedure

Treatments in this study were performed using a single-blinded, randomized, split-mouth design. Which quadrants received the conventional treatment (control quadrants) and which quadrants received experimental laser disinfection before SRP were selected in advance using a randomized computer-generated random table. The patients did not know which quadrants were to receive the active laser disinfection, since the control quadrants received inactive laser disinfection—where the laser tip was guided into the sulcus without being switched on. To exclude potential bias, the physicians who performed SRP and the measurements were also not aware of which quadrants were to receive the actual laser disinfection. This was ensured by having a separate practitioner performing the laser disinfection procedure (A.B.).

Before conventional pocket cleaning, the periodontal pockets in the experimental study group were disinfected with a diode laser (SiroLaser Blue, Dentsply Sirona, Bensheim, Germany). The wavelength was 445 nm and the power was 0.5 W, with the laser in continuous-wave mode (laser class 4). A 320 μm fiber was used to disinfect/clean circularly around the diseased tooth for 1 min under constant movement. The laser light used was guided into the gingival pocket with an optical fiber. The device setting of 0.5 W corresponded to an effective power of 0.48 W. Assuming a Gaussian laser beam profile, the power density was 1210 W/cm^2 (fiber tip diameter 0.32 mm; 0.48 W output power). Due to the low power density, this laser light is not harmful to the gums or other surrounding tissue. However, laser safety goggles must be worn during the irradiation procedure to protect the eyes.

All of the teeth with periodontal disease were then treated with manual instruments (Gracey curettes, Hu-Friedy, Leimen, Germany) and a piezoelectric scaler (Vector, Dürr

Dental, Bietigheim-Bissingen, Germany) with a slimline scaler tip. The treatment end point was the clinically assessed absence of concrements on the root surfaces. This was assessed by a dentist who was experienced in periodontal treatment (F.K.), who was not aware of which teeth had received active or inactive laser disinfection.

Figure 2 shows the sequence of individual steps in more detail.

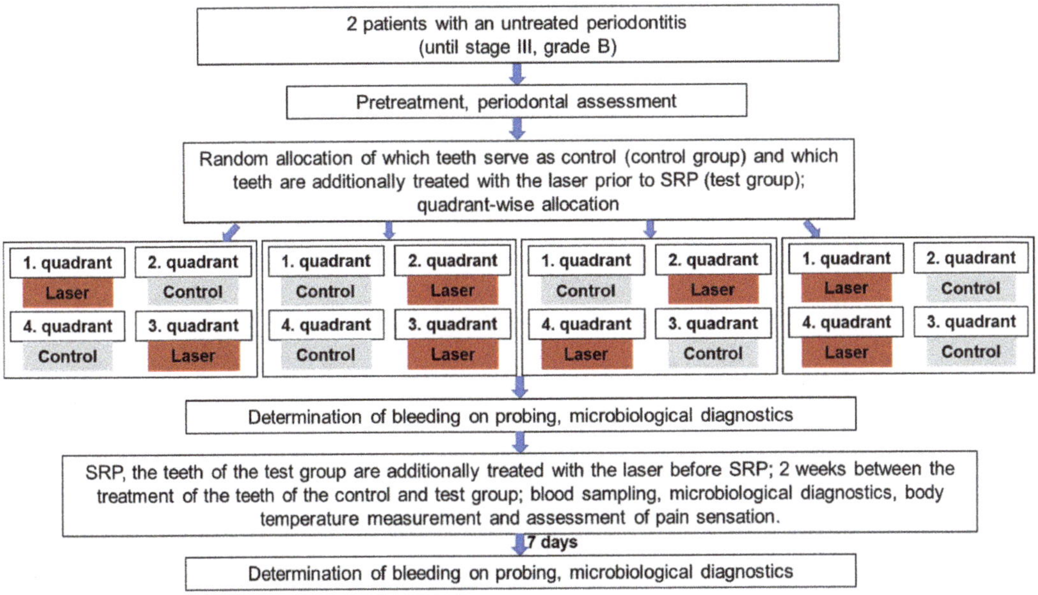

Figure 2. Flowchart for the study design.

4.4. Patients

Two patients (two men, one 39 years of age and the other 63) with untreated generalized periodontitis (stage III, grade B) were recruited in the Clinic of Operative Dentistry, Periodontology, and Preventive Dentistry at Aachen University Hospital. The inclusion criteria were a minimum age of 18 years, no periodontal treatment in the previous 2 years, generalized periodontitis up to stage III, grade B, and the provision of written informed consent. The exclusion criteria were patients who had already received periodontal treatment within the previous 2 years, pregnancy, tobacco use, dental implants, metabolic diseases with possible influence on the healing process (especially diabetes), infectious diseases (such as human immunodeficiency virus or hepatitis B or C), patients undergoing chemotherapy, cystic fibrosis, and antibiotic treatment within the previous 3 months. All patients received detailed information regarding the conduction of the study and data protection. Each of the participants gave written informed consent to participate in the study. In the test plan of the present study submitted to the local ethics committee, particular attention was paid to benefit–risk assessment, the additional study-related measures (blood sampling and laser disinfection), and the use of the laser within its approval. The local ethics committee of the RWTH Aachen University Hospital saw/sees no concerns regarding the research project from an ethical and professional perspective. The study was conducted in full accordance with local and global ethical guidelines (World Medical Association Declaration of Helsinki, version X, 2013) and approved by the local ethics committee (reference number: EK 197/21; date of approval 17 August 2021).

4.5. Statistical Analysis

In this feasibility study, descriptive statistical analysis methods were mainly used for the two patients included. Null hypothesis significance testing is not appropriate unless the sample size is properly powered [49].

5. Conclusions

This feasibility study showed that the study protocol appears to be suitable for detecting bacteremia during periodontal therapy. Initial indications of a bacteremia-reducing or even bacteremia-preventive effect of the laser disinfection method described were observed. However, due to the small number of patients, this effect must continue to be investigated and proven in follow-up studies with larger numbers of patients.

Author Contributions: Conceptualization, A.B.; methodology, P.J., G.C., and A.B.; formal analysis, P.J. and A.B.; investigation, P.J., J.-S.W., F.K., and A.B.; microbiological analyses, G.C.; writing—original draft preparation, P.J. and A.B.; writing—review and editing, P.J., G.C., J.-S.W., F.K., and A.B.; visualization, P.J.; project administration, P.J. and A.B. All authors have read and agreed to the published version of the manuscript.

Funding: This research received no external funding.

Institutional Review Board Statement: The study was conducted in accordance with the Declaration of Helsinki and approved by the Ethics Committee of the RWTH Aachen University Hospital (protocol code EK 197/21 and date of approval 17 August 2021).

Informed Consent Statement: Informed consent was obtained from all subjects involved in the study. Written informed consent was obtained from the patients to publish this paper.

Data Availability Statement: The data are not publicly available due to ethical restrictions regarding patient data.

Acknowledgments: We would like to thank Beate Melzer-Krick for her tireless assistance with the sample collection and microbiological testing. We would also like to thank the staff of the laboratory diagnostic center at RWTH Aachen University Hospital for the blood culture and immunological analyses.

Conflicts of Interest: The authors declare no conflict of interest.

References

1. Baeza, M.; Morales, A.; Cisterna, C.; Cavalla, F.; Jara, G.; Isamitt, Y.; Pino, P.; Gamonal, J. Effect of periodontal treatment in patients with periodontitis and diabetes: Systematic review and meta-analysis. *J. Appl. Oral. Sci.* **2020**, *28*, e20190248. [CrossRef]
2. Blaizot, A.; Vergnes, J.N.; Nuwwareh, S.; Amar, J.; Sixou, M. Periodontal diseases and cardiovascular events: Meta-analysis of observational studies. *Int. Dent. J.* **2009**, *59*, 197–209. [PubMed]
3. Manrique-Corredor, E.J.; Orozco-Beltran, D.; Lopez-Pineda, A.; Quesada, J.A.; Gil-Guillen, V.F.; Carratala-Munuera, C. Maternal periodontitis and preterm birth: Systematic review and meta-analysis. *Community Dent. Oral Epidemiol.* **2019**, *47*, 243–251. [CrossRef] [PubMed]
4. Mealey, B.L.; Ocampo, G.L. Diabetes mellitus and periodontal disease. *Periodontol. 2000* **2007**, *44*, 127–153. [CrossRef]
5. Preshaw, P.M.; Foster, N.; Taylor, J.J. Cross-susceptibility between periodontal disease and type 2 diabetes mellitus: An immunobiological perspective. *Periodontol. 2000* **2007**, *45*, 138–157. [CrossRef]
6. Salvi, G.E.; Carollo-Bittel, B.; Lang, N.P. Effects of diabetes mellitus on periodontal and peri-implant conditions: Update on associations and risks. *J. Clin. Periodontol.* **2008**, *35*, 398–409. [CrossRef] [PubMed]
7. Saremi, A.; Nelson, R.G.; Tulloch-Reid, M.; Hanson, R.L.; Sievers, M.L.; Taylor, G.W.; Shlossman, M.; Bennett, P.H.; Genco, R.; Knowler, W.C. Periodontal disease and mortality in type 2 diabetes. *Diabetes Care* **2005**, *28*, 27–32. [CrossRef] [PubMed]
8. Bahekar, A.A.; Singh, S.; Saha, S.; Molnar, J.; Arora, R. The prevalence and incidence of coronary heart disease is significantly increased in periodontitis: A meta-analysis. *Am. Heart J.* **2007**, *154*, 830–837. [CrossRef]
9. Mustapha, I.Z.; Debrey, S.; Oladubu, M.; Ugarte, R. Markers of systemic bacterial exposure in periodontal disease and cardiovascular disease risk: A systematic review and meta-analysis. *J. Periodontol.* **2007**, *78*, 2289–2302. [CrossRef]
10. Kaye, E.K.; Valencia, A.; Baba, N.; Spiro, A., 3rd; Dietrich, T.; Garcia, R.I. Tooth loss and periodontal disease predict poor cognitive function in older men. *J. Am. Geriatr. Soc.* **2010**, *58*, 713–718. [CrossRef]
11. Matsushita, K.; Yamada-Furukawa, M.; Kurosawa, M.; Shikama, Y. Periodontal Disease and Periodontal Disease-Related Bacteria Involved in the Pathogenesis of Alzheimer's Disease. *J. Inflamm. Res.* **2020**, *13*, 275–283. [CrossRef] [PubMed]

12. Noble, J.M.; Borrell, L.N.; Papapanou, P.N.; Elkind, M.S.; Scarmeas, N.; Wright, C.B. Periodontitis is associated with cognitive impairment among older adults: Analysis of NHANES-III. *J. Neurol. Neurosurg. Psychiatry* **2009**, *80*, 1206–1211. [CrossRef] [PubMed]
13. Stein, P.S.; Desrosiers, M.; Donegan, S.J.; Yepes, J.F.; Kryscio, R.J. Tooth loss, dementia and neuropathology in the Nun study. *J. Am. Dent. Assoc.* **2007**, *138*, 1314–1382. [CrossRef]
14. Ide, M.; Harris, M.; Stevens, A.; Sussams, R.; Hopkins, V.; Culliford, D.; Fuller, J.; Ibbett, P.; Raybould, R.; Thomas, R.; et al. Periodontitis and Cognitive Decline in Alzheimer's Disease. *PLoS ONE* **2016**, *11*, e0151081. [CrossRef] [PubMed]
15. Dominy, S.S.; Lynch, C.; Ermini, F.; Benedyk, M.; Marczyk, A.; Konradi, A.; Nguyen, M.; Haditsch, U.; Raha, D.; Griffin, C.; et al. *Porphyromonas gingivalis* in Alzheimer's disease brains: Evidence for disease causation and treatment with small-molecule inhibitors. *Sci. Adv.* **2019**, *5*, eaau3333. [CrossRef] [PubMed]
16. Singhrao, S.K.; Harding, A.; Poole, S.; Kesavalu, L.; Crean, S. *Porphyromonas gingivalis* Periodontal Infection and Its Putative Links with Alzheimer's Disease. *Mediat. Inflamm.* **2015**, *2015*, 137357. [CrossRef]
17. Haffajee, A.D.; Arguello, E.I.; Ximenez-Fyvie, L.A.; Socransky, S.S. Controlling the plaque biofilm. *Int. Dent. J.* **2003**, *53* (Suppl. 3), 191–199. [CrossRef]
18. Drisko, C.L.; Cochran, D.L.; Blieden, T.; Bouwsma, O.J.; Cohen, R.E.; Damoulis, P.; Fine, J.B.; Greenstein, G.; Hinrichs, J.; Somerman, M.J.; et al. Position paper: Sonic and ultrasonic scalers in periodontics. Research, Science and Therapy Committee of the American Academy of Periodontology. *J. Periodontol.* **2000**, *71*, 1792–1801. [CrossRef]
19. Tunkel, J.; Heinecke, A.; Flemmig, T.F. A systematic review of efficacy of machine-driven and manual subgingival debridement in the treatment of chronic periodontitis. *J. Clin. Periodontol.* **2002**, *29* (Suppl. S3), 72–91. [CrossRef]
20. Petersilka, G.J.; Tunkel, J.; Barakos, K.; Heinecke, A.; Haberlein, I.; Flemmig, T.F. Subgingival plaque removal at interdental sites using a low-abrasive air polishing powder. *J. Periodontol.* **2003**, *74*, 307–311. [CrossRef]
21. Schwarz, F.; Bieling, K.; Venghaus, S.; Sculean, A.; Jepsen, S.; Becker, J. Influence of fluorescence-controlled Er:YAG laser radiation, the Vector system and hand instruments on periodontally diseased root surfaces in vivo. *J. Clin. Periodontol.* **2006**, *33*, 200–208. [CrossRef] [PubMed]
22. Wenzler, J.S.; Krause, F.; Boecher, S.; Falk, W.; Birkenmaier, A.; Conrads, G.; Braun, A. Antimicrobial Impact of Different Air-Polishing Powders in a Subgingival Biofilm Model. *Antibiotics* **2021**, *10*, 1464. [CrossRef]
23. Coluzzi, D.J.; Mizutani, K.; Yukna, R.; Al-Falaki, R.; Lin, T.; Aoki, A. Surgical laser therapy for periodontal and peri-implant disease. *Clin. Dent. Rev.* **2022**, *6*, 7. [CrossRef]
24. Cobb, C.M. Lasers and the treatment of periodontitis: The essence and the noise. *Periodontol. 2000* **2017**, *75*, 205–295. [CrossRef]
25. Braun, A.; Krause, F.; Hartschen, V.; Falk, W.; Jepsen, S. Efficiency of the Vector -system compared with conventional subgingival debridement in vitro and in vivo. *J. Clin. Periodontol.* **2006**, *33*, 568–574. [CrossRef]
26. Eberhard, J.; Ehlers, H.; Falk, W.; Acil, Y.; Albers, H.K.; Jepsen, S. Efficacy of subgingival calculus removal with Er:YAG laser compared to mechanical debridement: An in situ study. *J. Clin. Periodontol.* **2003**, *30*, 511–518. [CrossRef] [PubMed]
27. Mombelli, A.; Schmid, B.; Rutar, A.; Lang, N.P. Local antibiotic therapy guided by microbiological diagnosis. *J. Clin. Periodontol.* **2002**, *29*, 743–749. [CrossRef] [PubMed]
28. Kim, T.S.; Burklin, T.; Schacher, B.; Ratka-Krueger, P.; Schaecken, M.T.; Renggli, H.H.; Fiehn, W.; Eickholz, P. Pharmacokinetic profile of a locally administered doxycycline gel in crevicular fluid, blood, and saliva. *J. Periodontol.* **2002**, *73*, 1285–1291. [CrossRef]
29. Patianna, G.; Valente, N.A.; D'Addona, A.; Andreana, S. In vitro evaluation of controlled-release 14% doxycycline gel for decontamination of machined and sandblasted acid-etched implants. *J. Periodontol.* **2018**, *89*, 325–330. [CrossRef]
30. Sanz, M.; Herrera, D.; Kebschull, M.; Chapple, I.; Jepsen, S.; Beglundh, T.; Sculean, A.; Tonetti, M.S.; Participants, E.F.P.W.; Methodological, C. Treatment of stage I-III periodontitis-The EFP S3 level clinical practice guideline. *J. Clin. Periodontol.* **2020**, *47* (Suppl. 22), 4–60. [CrossRef]
31. Forner, L.; Larsen, T.; Kilian, M.; Holmstrup, P. Incidence of bacteremia after chewing, tooth brushing and scaling in individuals with periodontal inflammation. *J. Clin. Periodontol.* **2006**, *33*, 401–407. [CrossRef] [PubMed]
32. Horliana, A.C.; Chambrone, L.; Foz, A.M.; Artese, H.P.; Rabelo Mde, S.; Pannuti, C.M.; Romito, G.A. Dissemination of periodontal pathogens in the bloodstream after periodontal procedures: A systematic review. *PLoS ONE* **2014**, *9*, e98271. [CrossRef] [PubMed]
33. Kinane, D.F.; Riggio, M.P.; Walker, K.F.; MacKenzie, D.; Shearer, B. Bacteraemia following periodontal procedures. *J. Clin. Periodontol.* **2005**, *32*, 708–713. [CrossRef] [PubMed]
34. Hollingshead, C.M.; Brizuela, M. Antibiotic Prophylaxis in Dental and Oral Surgery Practice. In *StatPearls*; StatPearls Publishing: Treasure Island, FL, USA, 2023.
35. Caruso, U.; Nastri, L.; Piccolomini, R.; d'Ercole, S.; Mazza, C.; Guida, L. Use of diode laser 980 nm as adjunctive therapy in the treatment of chronic periodontitis. A randomized controlled clinical trial. *New Microbiol.* **2008**, *31*, 513–518. [PubMed]
36. Zhao, P.; Song, X.; Wang, Q.; Zhang, P.; Nie, L.; Ding, Y.; Wang, Q. Effect of adjunctive diode laser in the non-surgical periodontal treatment in patients with diabetes mellitus: A systematic review and meta-analysis. *Lasers Med. Sci.* **2021**, *36*, 939–950. [CrossRef]
37. Assaf, M.; Yilmaz, S.; Kuru, B.; Ipci, S.D.; Noyun, U.; Kadir, T. Effect of the diode laser on bacteremia associated with dental ultrasonic scaling: A clinical and microbiological study. *Photomed. Laser Surg.* **2007**, *25*, 250–256. [CrossRef]
38. Kornman, K.S.; Crane, A.; Wang, H.Y.; di Giovine, F.S.; Newman, M.G.; Pirk, F.W.; Wilson, T.G., Jr.; Higginbottom, F.L.; Duff, G.W. The interleukin-1 genotype as a severity factor in adult periodontal disease. *J. Clin. Periodontol.* **1997**, *24*, 72–77. [CrossRef]

39. Serrage, H.; Heiskanen, V.; Palin, W.M.; Cooper, P.R.; Milward, M.R.; Hadis, M.; Hamblin, M.R. Under the spotlight: Mechanisms of photobiomodulation concentrating on blue and green light. *Photochem. Photobiol. Sci.* **2019**, *18*, 1877–1909. [CrossRef]
40. Schaefer, B. The Effect of Dental Laser Systems on Periodontal Pathogenic Microorganisms In Vitro. Bachelor's Thesis, Christian-Albrechts-Universität zu Kiel, Kiel, Germany, 2015.
41. Lafaurie, G.I.; Mayorga-Fayad, I.; Torres, M.F.; Castillo, D.M.; Aya, M.R.; Baron, A.; Hurtado, P.A. Periodontopathic microorganisms in peripheric blood after scaling and root planing. *J. Clin. Periodontol.* **2007**, *34*, 873–879. [CrossRef]
42. Zhang, W.; Daly, C.G.; Mitchell, D.; Curtis, B. Incidence and magnitude of bacteraemia caused by flossing and by scaling and root planing. *J. Clin. Periodontol.* **2013**, *40*, 41–52. [CrossRef]
43. Beutler, J.; Jentsch, H.F.R.; Rodloff, A.C.; Stingu, C.S. Bacteremia after professional mechanical plaque removal in patients with chronic periodontitis. *Oral Dis.* **2019**, *25*, 1185–1194. [CrossRef]
44. Graziani, F.; Cei, S.; Orlandi, M.; Gennai, S.; Gabriele, M.; Filice, N.; Nisi, M.; D'Aiuto, F. Acute-phase response following full-mouth versus quadrant non-surgical periodontal treatment: A randomized clinical trial. *J. Clin. Periodontol.* **2015**, *42*, 843–852. [CrossRef] [PubMed]
45. Geerts, S.O.; Nys, M.; De, M.P.; Charpentier, J.; Albert, A.; Legrand, V.; Rompen, E.H. Systemic release of endotoxins induced by gentle mastication: Association with periodontitis severity. *J. Periodontol.* **2002**, *73*, 73–78. [CrossRef] [PubMed]
46. Thornhill, M.H.; Dayer, M.; Lockhart, P.B.; Prendergast, B. Antibiotic Prophylaxis of Infective Endocarditis. *Curr. Infect. Dis. Rep.* **2017**, *19*, 9. [CrossRef] [PubMed]
47. Balejo, R.D.P.; Cortelli, J.R.; Costa, F.O.; Cyrino, R.M.; Aquino, D.R.; Cogo-Muller, K.; Miranda, T.B.; Moura, S.P.; Cortelli, S.C. Effects of chlorhexidine preprocedural rinse on bacteremia in periodontal patients: A randomized clinical trial. *J. Appl. Oral Sci.* **2017**, *25*, 586–595. [CrossRef] [PubMed]
48. Poole, S.; Singhrao, S.K.; Kesavalu, L.; Curtis, M.A.; Crean, S. Determining the presence of periodontopathic virulence factors in short-term postmortem Alzheimer's disease brain tissue. *J. Alzheimers Dis.* **2013**, *36*, 665–677. [CrossRef] [PubMed]
49. Tickle-Degnen, L. Nuts and bolts of conducting feasibility studies. *Am. J. Occup. Ther.* **2013**, *67*, 171–176. [CrossRef]

antibiotics

MDPI

Article

Matrix Metalloproteinase 9 (MMP-9) and Interleukin-8 (IL-8) in Gingival Crevicular Fluid after Minimally Invasive Periodontal Surgery with or without Er:YAG and Nd:YAG Laser Application

Ewa Dolińska [1,*], Anna Skurska [1,2], Violetta Dymicka-Piekarska [3], Robert Milewski [4] and Małgorzata Pietruska [1]

1 Department of Periodontal and Oral Mucosa Diseases, Medical University of Bialystok, ul. Waszyngtona 13, 15-269 Bialystok, Poland; annaskurska@wp.pl (A.S.); mpietruska@wp.pl (M.P.)
2 Department of Integrated Dentistry, Medical University of Bialystok, ul. M. Skłodowskiej-Curie 24A, 15-276 Bialystok, Poland
3 Department of Clinical Laboratory Diagnostics, Medical University of Bialystok, ul. Waszyngtona 15, 15-269 Bialystok, Poland; violetta.dymicka-piekarska@umb.edu.pl
4 Department of Biostatistics and Medical Informatics, Medical University of Bialystok, ul. Szpitalna 37, 15-295 Bialystok, Poland; robert.milewski@umb.edu.pl
* Correspondence: ewa.dolinska@umb.edu.pl; Tel.: +48-85-748-59-05

Abstract: Background: This study aimed to evaluate alterations in the concentrations of matrix metalloproteinase-9 (MMP-9) and interleukin-8 (IL-8) within gingival crevicular fluid (GCF) extracted from the intrabony periodontal defect site before and after minimally invasive regenerative surgery, with or without supplemental laser application. The surgical procedure was performed using the modified minimally invasive surgical technique (M-MIST). Methods: Thirty-eight patients, each presenting with a single vertical defect, were randomly assigned to either the test (M-MIST + Er:YAG + Nd:YAG) or the control group (M-MIST). IL-8 and MMP-9 levels (primary outcomes of the study) were assessed prior to therapy, after 2 and 4 weeks, and 6 months following the surgical procedure by means of dedicated ELISA kits. Results: Both procedures were clinically effective as evidenced by probing depth (PD) reduction and clinical attachment level (CAL) gain at the 6-month follow-up. No statistical differences were observed in the levels of MMP-9 and IL-8 between the groups at any time point assessed. The changes in the level of MMP-9 and IL-8 over time were not statistically significant in any group. IL-8 was positively correlated with MMP-9 in the control group throughout the study and in the test group 2 weeks and 6 months post-op. Conclusions: Within the limitations of this study, the additional application of Er:YAG + Nd:YAG lasers alongside the M-MIST procedure did not enhance the clinical and biochemical treatment outcomes compared to M-MIST alone.

Keywords: periodontal regeneration; interleukin-8; metalloproteinase-9; inflammatory markers; periodontal intrabony defects; Er:YAG laser; Nd:YAG laser

1. Introduction

Periodontitis is a multifactorial, infectious disease that leads to the immunologically mediated damage of teeth-supporting tissues and, consequently, tooth loss [1]. The course of periodontitis varies among patients and its advanced forms concern about 10 percent of the population worldwide, which makes it a social disease [2]. The wide spread of the disease means that its treatment generates tremendous costs for healthcare systems worldwide [3].

In some patients, in the course of periodontitis, intrabony defects occur. Their etiology is not fully explained yet. Occlusal trauma, food impaction, or the presence of plaque-retention factors are taken into account in their formation. Attention is also paid to the distance between the roots of adjacent teeth [4]. In the case of root proximity, inflammation affects the entire bone septum between roots. If the bone is wider than the "firing" range of dental plaque, the bone septum is only partially destroyed, which results in the formation of a subcrestal bone defect [5]. The presence of intrabony defects is indicative of stage III and IV periodontitis, thereby exacerbating the prognosis of the disease [6].

The presence of intrabony defects is an indication for regenerative procedures. The most frequently used are guided tissue regeneration (GTR) and biomodification of the root with enamel matrix derivatives (EMDs) or EMD and bone graft [7–9]. Nowadays, more and more often, minimally invasive surgery (MIS) in periodontal regeneration is performed. MIS offers the advantage of reducing morbidity, minimizing scars, and mitigating other postsurgical complications in treated patients [10]. In cases of bone defects that spread no more than one-third of the circumference of the tooth, there is a possibility to use an M-MIST (modified minimally invasive surgical technique) surgery, which shows promising results [11]. However, a previous study reported no statistically significant differences in terms of CAL gain and PD reduction when MIS was compared with an extended flap with papilla preservation [12].

Laserotherapy's importance has also been increasing in recent years. It is used as an adjunct to periodontal treatment [13,14]. The Er:YAG laser has the ability to ablate hard and soft tissues [15]. It can also modify the root surface and remove dental calculus [16,17]. Er:YAG laser application facilitates the adhesion of fibroblasts to the root surface better than scaling and root planing alone [18]. The Nd:YAG laser has different indications. It is used to inactivate microorganisms, to remove the periodontal pocket epithelium, and to improve hemostasis [19]. There are also reports of the combined use of both types of lasers in periodontal non-surgical treatment. The Er:YAG laser enables decontamination of the root surface and hard deposits ablation, and the Nd:YAG laser accelerates clot formation [20,21].

Non-resolving inflammation is a main cause of periodontal hard and soft tissue destruction [22]. Cytokines, matrix metalloproteinases (MMPs), and their inhibitors, growth factors and other regulators in the periodontium, are synthesized in response to bacteria and their products, inducing and maintaining the inflammatory response [23]. IL-8 is a chemoattractant cytokine produced by a variety of tissue and blood cells with a distinct target for recruitment and activation of neutrophils [24]. Neutrophils represent the major inflammatory infiltrate in the periodontium. Neutrophil enzymes released after activation can effectively degrade connective tissue. IL-8 is of considerable interest for a better understanding of the mechanisms leading to neutrophil connecting tissue destruction [25]. Wide range of metalloproteinases (MMPs), enzymes that degrade the extracellular matrix within periodontal tissues. are also involved in this process. MMPs' expression in mature tissues is normally low, but it rises in pathologies like inflammatory diseases, metastasis, or tumor growth [26]. MMPs are a big family of zinc-dependent endopeptidases and collectively are able to destroy most of the extracellular matrix proteins in the periodontium. Many studies have described interstitial collagenases (collagenases 1, 2, 3, MMP-1, MMP-8, MMP-13), gelatinases A and B (MMP-2, MMP-9) and stromelysin (MMP-3) in gingival crevicular fluid (GCF) [27]. However, interpreting the GCF concentration of specific enzymes is challenging. For instance, enzymes derived from neutrophils, like MMP-8, offer information about the number of these cells in the periodontal pocket rather than the destruction of periodontal tissues. The role of gelatinases in inflammation is also still not clear, and its GCF concentration is hard to interpret [28]. A relationship between MMP-9 GCF concentration and clinical periodontal parameters including attachment loss was noted [29]. It was also found that levels of MMP-9 are increased in patients with periodontitis and decrease after periodontal therapy [30]. MMP-9 is involved not only in the soft tissue breakdown, but there are scarce reports of its indirect role in immune-mediated bone loss in the periodontium [31,32], so MMP-9 may play a role in the mediating of the bone defects' formation.

Given the aforementioned considerations, the objective of our study was to assess changes in the levels of MMP-9 and IL-8 in periodontal pockets corresponding to intrabony periodontal defects eligible for regenerative procedures, specifically M-MIST alone or M-MIST with additional Er:YAG and Nd:YAG laser applications, over a six-month follow-up period. We assumed changes in immunological parameters (MMP-9, IL-8) as the primary endpoints and clinical parameters, such as PD reduction and CAL gain, as the secondary endpoints of the study. We formulated the null hypothesis that additional laser usage does not affect MMP-9 and IL-8 levels, nor does it affect the values of clinical parameters.

2. Results

All 38 patients with periodontitis (stage III) [33] completed the 6-month follow-up visits with no further dropouts and a 100% retention rate. No adverse events were reported. Patients were equally distributed between the study and control groups without differences according to age, gender, or intrasurgical defect depth. Table 1 depicts the characteristics of patients, with teeth and defects included.

Table 1. Characteristics of study participants and teeth included in the surgical procedures (clinical parameters in mm).

	Test (M-MIST + Er:YAG + Nd:YAG)	Control (M-MIST)	Significance
Number of patients	19	19	-
Gender	12F/7M	9F/10M	NS (p * = 0.34)
Age (range)	47 (30–73)	43.5 (24–59)	NS (p * = 0.9)
Incisors/canines/premolars/molars	3/3/12/1	3/1/9/6	-
Mean intrasurgical defect depth	4.03 ± 1.58	4.34 ± 1.13	NS (p * = 0.14)
FMPS (%)	10.8 ± 4.42	11.6 ± 4.68	NS (p * = 0.53)
FMBOP (%)	10.71 ± 3.89	10.34 ± 4.58	NS (p * = 0.83)

p *—U Mann–Whitney test.

Included patients were characterized by good oral hygiene expressed by the FMPS (full mouth plaque score) index [34]. At baseline, there was no person with an FMPS over 20%. The mean FMPS in the test group amounted to 10.8%, and it was 11.6% in the control group. FMBOP (full mouth bleeding on probing) was also low [35]. It was 10.7% in the test group and 10.3% in the control group, respectively.

Clinical parameters of the operated area are presented in Table 2 and Figure 1. In both groups, there were significant PD reduction and CAL gain in the 6-month observation period. Gingival recession (GR) increased in the control group and did not change in the test group. The changes in GR were not significant over time. There were also no significant differences between both groups studied before or six months post-op.

Table 2. Clinical parameters in test (M-MIST + Er:YAG) and control (M-MIST) groups at baseline and 6 months post-op.

	PD (mm)			GR (mm)			CAL (mm)		
	M-MIST + Er:YAG + Nd:YAG	M-MIST	p **	M-MIST + Er:YAG + Nd:YAG	M-MIST	p **	M-MIST + Er:YAG + Nd:YAG	M-MIST	p **
Baseline (mean)	7.62 ± 1.44	7.15 ± 1.25		1.31 ± 1.34	0.89 ± 1.19		8.57 ± 2.16	8.05 ± 1.8	
6 months (mean)	4.42 ± 1.30	4.18 ± 1.14	NS	1.31 ± 1.00	1.13 ± 1.35	NS	5.73 ± 1.48	5.31 ± 2.02	NS
	p * = 0.0001	p * = 0.0002		NS	NS		p * = 0.0002	p * = 0.0003	
Baseline (median)	7 (6–11)	7 (5–9)		1.5 (0–4)	0 (0–4)		8 (5–15)	8 (5–12)	
6 months (median)	4 (3–7)	4 (3–7)		1 (0–3)	1 (0–3)		6 (4–9)	5 (3–10)	
Diff.	2.84 ± 0.96	2.97 ± 1.18	NS	0.00 ± 0.86	−0.24 ± 0.79	NS	2.84 ± 1.45	2.74 ± 1.45	NS

p *—Wilcoxon pair test (changes in time); p **—U Mann–Whitney test (between groups); NS—non-significant; Diff.—difference 0–6.

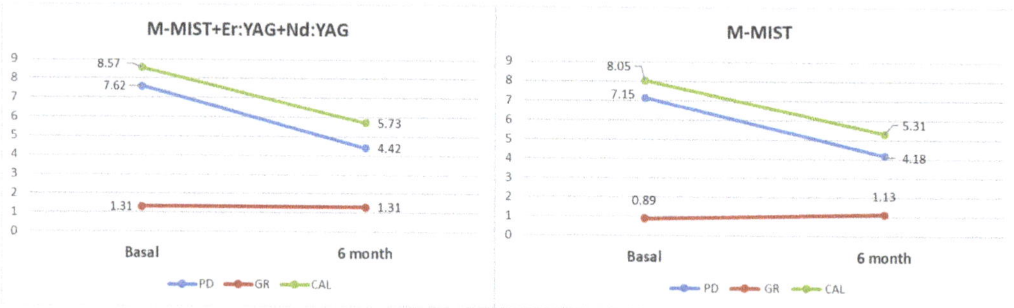

Figure 1. Mean changes in clinical parameters at the six-month follow-up in the test and control groups.

Changes in MMP-9 levels (Table 3, Figure 2a) obtained from periodontal pockets that underwent surgical treatment were not significant in any group at the six-month follow-up. We did not note differences between groups either. However, after Er:YAG and Nd:YAG laser application, the MMP-9 amount increased much more in the test than in the control group after 2 and 4 weeks (without significance).

Table 3. Mean changes in the MMP-9 amount at the six-month follow-up in the test (M-MIST + Er:YAG + Nd:YAG) and control (M-MIST) groups expressed as ng/mL per 30 s sample.

	MMP-9		
	M-MIST + Er:YAG + Nd:YAG	**M-MIST**	**p ** (Between Groups)**
Baseline	61.96 ± 87.86	42.32 ± 30.52	NS
2 weeks	109.04 ± 207.35	57.51 ± 65.86	NS
4 weeks	115.96 ± 125.99	53.34 ± 46.71	NS
6 months	59.78 ± 59.45	58.57 ± 49.44	NS
p * (changes in time)	NS	NS	
Diff. 0–2 w	−47.08 ± 187.36	−15.19 ± 70.34	NS
Diff. 0–4 w	−54 ± 151.00	−11.02 ± 54.02	NS
Diff. 0–6 m	2.18 ± 89.74	−16.25 ± 52.14	NS

p *—Anova Friedman's for multiple comparisons; p **—U Mann–Whitney test; NS—non-significant, Diff.—difference.

Similar results for the IL-8 amount were noted. There were no significant differences between groups or at the 6-month follow-up in any group. But, the most visible changes in IL-8 levels were noted 2 and 4 weeks after Er:YAG and Nd:YAG laser surgery, although this was without significance. Mean changes in IL-8 levels are depicted in Table 4 and Figure 2b.

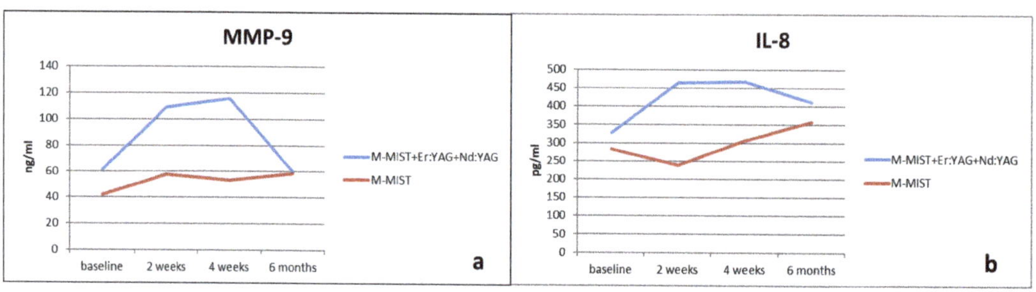

Figure 2. (**a**) Mean changes in the MMP-9 amount at the six-month follow-up. (**b**) Mean changes in the IL-8 amount at the six-month follow-up.

Table 4. Mean changes in the IL-8 amount at the six-month follow-up in the test (M-MIST + Er:YAG + Nd:YAG) and control (M-MIST) groups expressed as pg/mL per 30 s sample.

	IL-8		
	M-MIST + Er:YAG + Nd:YAG	**M-MIST**	***p* ** (Between Groups)**
Baseline	327.22 ± 281.55	283.20 ± 158.41	NS
2 weeks	464.84 ± 477.98	238.03 ± 141.13	NS
4 weeks	467.91 ± 313.12	307.77 ± 256.33	NS
6 months	410.97 ± 300.62	357.77 ± 229.54	NS
p * (changes in time)	NS	NS	
Diff. 0–2 w	−137.62± 448.06	45.17± 182.89	NS
Diff. 0–4 w	−140.69± 387.93	−24.57± 311.88	NS
Diff. 0–6 m	−83.75 ± 341.86	−74.56 ± 285.53	NS

p *—Anova Friedman's for multiple comparisons; *p* **—U Mann–Whitney test; NS—non-significant, Diff.—difference.

The GCF volume (SFFR, sulcus fluid flow rate) expressed in relative PU did not change over time in the test group, but it was on the verge of significance in the control group during the six-month observation period. There were no differences between groups. The SFFR in both groups slightly increased after a week following post-surgical intervention to increase after 2 weeks and finally decrease after 6 months. Mean changes in the SFFR are shown in Table 5 and Figure 3.

Table 5. Mean changes in the SFFR volume expressed in the relative PU at the six-month follow-up in the test (M-MIST + Er:YAG + Nd:YAG) and control (M-MIST) groups per 30 s sample.

	SFFR		
	M-MIST + Er:YAG + Nd:YAG	**M-MIST**	***p* ** (Between Groups)**
Baseline	98.89 ± 30.14	98.42 ± 43.21	NS
2 weeks	109.31 ± 33.65	102.00 ± 37.08	NS
4 weeks	77.15 ± 28.67	75.78 ± 30.75	NS
6 months	95.57 ± 46.96	87.00 ± 35.46	NS
p * (changes in time)	NS	$p = 0.046$	
Diff. 0–2 w	−10.42 ± 53.56	−3.58 ± 53.40	NS
Diff. 0–4 w	21.74± 37.33	22.64 ± 46.15	NS
Diff. 0–6 m	3.32 ± 43.13	11.42 ± 51.02	NS

p *—Anova Friedman's for multiple comparisons; *p* **—U Mann–Whitney test; NS—non-significant, Diff.—difference.

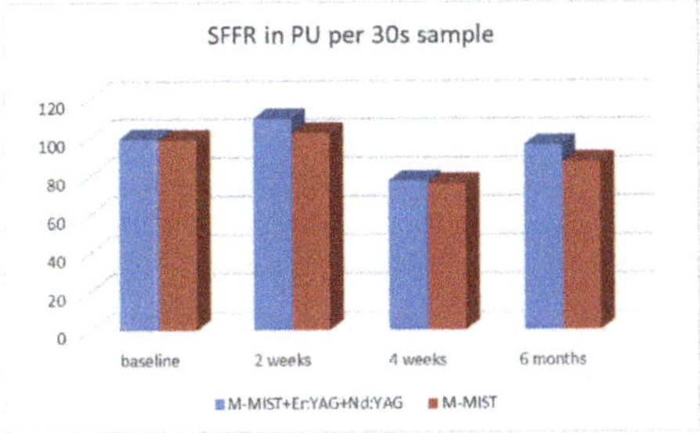

Figure 3. Mean changes in the SFFR volume in the 30 s sample at the six-month follow-up in the test and control groups.

The most important correlation noted was the moderate dependence of IL-8 and MMP-9. The amount of both molecules correlated at every time point in the control group. Similar correlations were observed in the test group, but only after 2 weeks and 6 months. Many other correlations were checked. MMP-9 correlated with SFFR after 4 weeks and 6 months, and MMP-9 correlated with PD after six months in the control group. MMP-9 and SFFR correlated in the test group only after 6 months. Spearman rank order correlations are shown in Tables 6 and 7.

Table 6. Correlations (Spearman test) between GCF IL-8 and MMP-9 levels at the 6-month follow-up in the test (M-MIST + Er:YAG + Nd:YAG) and control (M-MIST) groups in the operated region.

IL-8 and MMP-9	M-MIST + Er:YAG + Nd:YAG		M-MIST	
	R	*p*	R	*p*
Baseline	-	NS	0.62	0.0045
2 weeks	0.68	0.0015	0.74	0.0003
4 weeks	-	NS	0.59	0.0082
6 months	0.58	0.0087	0.67	0.0016
NS—non-significant				

Table 7. Other investigated correlations (Spearman test) in the test (M-MIST + Er:YAG + Nd:YAG) and control (M-MIST) groups.

	Correlation	M-MIST + Er:YAG + Nd:YAG	M-MIST
Baseline	IL-8 and PD	NS	NS
Baseline	IL-8 and SSFR	NS	NS
Baseline	IL-8 and intra-defect depth	NS	NS
Baseline	MMP-9 and PD	NS	NS
Baseline	MMP-9 and SSFR	NS	NS
Baseline	MMP-9 and intra-defect depth	NS	NS
2 weeks	IL-8 and SFFR	NS	NS
2 weeks	MMP-9 and SFFR	NS	NS
4 weeks	IL-8 and SFFR	NS	NS
4 weeks	MMP-9 and SFFR	NS	R = 0.5, p = 0.03
6 months	IL-8 and PD	NS	NS
6 months	IL-8 and SSFR	NS	NS
6 months	MMP-9 and PD	NS	R = 0.5, p = 0.026
6 months	MMP-9 and SSFR	R = 0.46, p = 0.047	R = 0.5, p = 0.029
NS—non-significant			

3. Discussion

The results of the present study have shown that the effectiveness of intrabony defect treatment with M-MIST or M-MIST in combination with Er:YAG + Nd:YAG lasers is similar. Both methods were equally effective and led to statistically significant improvements in clinical parameters. PD reduction and CAL gain were almost the same in both groups, with a slight difference in favor of the M-MIST group (without significance). Healing was uneventful in both groups, and no adverse events were noted. This proves that the procedures performed were not invasive and the post-treatment healing was good. The results that plaque and bleeding indices were low indicate proper hygienic preparation of the patients and an optimal level of plaque control before reconstructive surgery, which is essential for undisturbed healing. Another important factor influencing the response to periodontal surgical treatment is smoking. Smoking has a negative effect on bone healing and regeneration [36]. That is why smokers were not included in the study.

In the case of the coexistence of residual pockets with vertical defects, periodontal reconstructive surgery procedures are always considered [37]. Recognized factors that also influence the effect of surgery include obtaining appropriate decontamination of the treatment area. Additional laser usage as a new solution may be a valuable option in the standard surgical procedure. By eliminating calculus and bacteria, Erbium-Doped Yttrium Aluminium Garnet (Er:YAG) and Neodymium-Doped Yttrium Aluminum Garnet (Nd:YAG) lasers can create a biocompatible root surface that may facilitate periodontal healing [38]. Due to different features, these two lasers have different effects on the hard and soft tissues. However, it should be taken into consideration that the use of lasers in periodontal treatment gives ambiguous results [14]. There are not many papers published evaluating the combined utilization of Er:YAG + Nd:YAG lasers in oral surgery. The available literature mostly refers to the use of Er:YAG + Nd:YAG in non-surgical periodontal treatment, where such a combination of lasers may lead to additional clinical improvements compared to non-surgical treatment alone [21].

Clinical examination is a gold standard to evaluate the effectiveness of periodontal reconstructive procedures, but knowledge about the changes that occur at the molecular level in the treated site are of most interest. Not much is still known about gingival crevicular fluid (GCF) immunological changes in intrabony defects after regenerative procedures [39]. We assessed the levels of IL-8 and MMP-9 in GCF obtained from the periodontal pockets corresponding to vertical defects subjected to M-MIST or M-MIST with additional laser utilization.

GCF is an altered serum transudate/inflammatory exudate. In healthy tissues, GCF originates from blood vessels, and it is produced thanks to the osmotic gradient. However, leukocytic infiltrate and polymorphonuclear leukocytes are always present in the junctional epithelium and in the sulcus. Numerous cytokines and enzymes are released to GCF, including host response factors, molecules from blood, local tissues, and plaque bacteria [40]. The fluid is easily collected, so it is a valuable source of locally and systematically derived biomarkers. It is at the center of contemporary research projects because useful diagnostic biological markers are researched to detect subclinical alterations in tissue metabolism before clinically visible damage can occur. As it is known, not only the composition but also the volume and flow rate of GCF are of importance. The relationship between the enlarged volume of GCF and the increased severity of inflammation has been well-documented [41,42]. In our study, the sulcus flowing flow rate (SFFR) increased at 2 weeks after surgery to decrease at 4 weeks and thereafter come back to baseline levels at 6 months. This is in line with the observations of other authors, who observed a rise in the volume of GCF shortly after surgery (0–2 weeks) and thereafter a volume decrease [43,44]. In previous work, our group investigated molecular changes in GCF after periodontal regenerative procedures, and we also noticed a rise in the SFFR one week after the surgery, and it remained high 2 weeks following the procedure. This post-surgical increase in GCF volume may indicate the enhancement of inflammation during early surgical wound healing [45].

Intrabony periodontal defects may differ from healthy sites in terms of the molecular GCF profile. This profile may also be changed by periodontal treatment. Not much is known about such changes, but the first studies trying to find immunological descriptions of intrabony defects have been recently published. In the clinical study comparing the molecular profile of intrabony defects with healthy sites utilizing a multiplex bead immunoassay, 27 markers were assessed. Sites with an intrabony defect presented increased IL-1α, IL-1β, IL-6, INFγ, and MMP-8 levels compared with healthy sites. Additionally, FGF and VEGF levels were elevated, and traits of cell aging were observed [46]. The same group examined the molecular profile of periodontal pockets corresponding to intrabony defects after minimally invasive non-surgical periodontal therapy (MINST) [47]. The authors focused on changes after 1 and 5 days and 3 months after MINST. The most pronounced molecular changes were observed a day after treatment. It returned to baseline after

3 months. Levels of IL-2, IL-4, IL-8, MMP-1, MMP-3, TIMP-1, and FGFb had significantly risen in GCF one day after non-surgical periodontal therapy.

Not many scientific studies refer to immunological parameters of surgically treated sites in the periodontium. Pellegrini et al. assessed wound healing proteins in GCF after regenerative and open flap debridement procedures. They concluded that MMP-1 and bone morphogenetic protein-7 were associated with periodontal regeneration because of the increase in its levels in patients who responded well to therapy [48]. Accelerated healing of infrabony defects after enamel matrix derivative application expressed by a rapid return to baseline of TIMP-1, MMP-1, and MMP-8 levels was observed by Okuda et al. [44]. Other authors found transforming growth factor-β1 useful for monitoring periodontal repair and regeneration [43] or the expression of growth mediators in GCF of patients after periodontal surgery [49].

We opted to evaluate MMP-9 and IL-8 due to existing evidence in the literature suggesting that these molecules may play a role not only in the destruction of soft tissues but also in bone resorption and remodeling. MMPs are proteases involved in extracellular matrix destruction in periodontitis. MMP-9 is responsible for the degradation of many proteins, such as basement membrane collagen and laminin [28]. Elevated levels of MMP-9 were observed in periodontitis patients compared with healthy subjects. It was demonstrated that periodontal treatment reduced the MMP-9 level dramatically [50,51]. There are many scientific studies assessing the collagenolytic role of MMPs, but there is still a scarcity of data from studying MMPs as regulators of periodontal inflammation [51]. It was proven that MMP-9 along with MMP-13 are involved in alveolar bone resorption and periodontal tissue destruction [52,53]. MMPs may not only degrade bone collagen matrix but also may indirectly modulate bone resorption via osteoclast activation. The mechanisms include the activation of osteoclast-secreted proMMP-9, which denaturates collagen derived from MMP-13 activity. The other is cleaving galectin-3, which is an inhibitor of osteoclastogenesis that nullifies its inhibitory effect. The next is regulating the receptor activator of nuclear factor-κB ligand (RANKL)/osteoprotegrin axis in favor of RANKL [31,51]. Taking into consideration the above, it can be assumed that MMP-9 is a regulator of periodontal bone lesions. We observed no significant differences between control and study groups according to MMP-9 levels. There were also no intragroup changes over time according to gelatinase B. However, in both groups, early after surgery (2 and 4 weeks), the amount of MMP-9 had risen. This change was more pronounced in the study group, but it was not significant. In the test group, there was a sharp increase, and, evidently, this change was more pronounced in this group but it was not significant. Interestingly, after 6 months, in both groups the level of MMP-9 came back to the value that was near the baseline level, and it was almost the same in both groups. This may indicate an increase in the release of MMP-9 after laser stimulation. Unfortunately, this cannot be confirmed because no study assessed MMP-9 after surgical therapy with additional laser usage.

Sijari et al. assessed IL-8 after resective periodontal surgery. They observed a decrease in its levels after therapy along with better early healing [54]. This is partially consistent with our results, because in the M-MIST group after 2 weeks, the amount of IL-8 was decreased, but then it rose to be higher than baseline levels. In the test group, we noticed an increase in IL-8 at all time points examined. Our results were not significant. There were no differences between groups either. What is worth noting after laser usage is that there was an increase in the IL-8 level in the test group, and there was a horizontal trend in its high concentration lasting 2 weeks. Then, after 6 months, the IL-8 level slightly decreased. In contrast, in the control group, right after surgery the IL-8 amount decreased, and after 6 months it reached higher than baseline levels. Despite the lack of statistical differences, this proves that laser therapy has an impact on the IL-8 level.

IL-8 (CXCL8) is a strong chemoattractant for neutrophils, and it stimulates fast degranulation of gelatinase B. MMP-9 cleaves IL-8 to more potent chemoattractants. It results in a positive feedback loop for neutrophil activation and chemotaxis and in the increased influx of neutrophils to fight infections [55]. In our study, the level of IL-8 correlated positively with MMP-9 at all time points in the control group and after 2 weeks and 6 months in the study group. The strength of correlations was moderate. It is interesting that despite the many parameters checked we recorded only single additional relationships between them, i.e., MMP-9 correlated positively with SFFR after 6 months in both groups. Our research group observed a similar positive correlation between IL-8 and MMP-9 after guided tissue regeneration with and without systemic antibiotics throughout the study [45].

Our study is well-planned and meticulously conducted research, but it has some drawbacks. A major limitation was the mediocre clinical sample size. This sample size was calculated on the basis of trials investigating clinical parameters after periodontal reconstructive surgery. Recruiting more patients would allow for a larger sample for biochemical evaluation. A weak point as well is the lack of reference values for immunological parameters, as the cytokine profile after different treatment strategies is only investigated. There is still a lack of knowledge of many molecular processes occurring in the periodontium in the state of inflammatory disease and after surgical and non-surgical therapies. Another aspect is not including smokers, and, therefore, these results should be taken with caution as smokers will undergo these kinds of clinical interventions.

Innovative procedures very often bring additional benefits and become standard treatment options. However, this is not always the case. In our study, we fail to find any advantages of laser-assisted M-MIST surgery. Neither clinical nor molecular results confirmed the superiority of the Er:YAG or Nd:YAG procedure. However, more clinical studies should be conducted to explore the potential of Er:YAG and Nd:YAG lasers in periodontal surgical therapy. The role of MMP-9 and IL-8 in periodontal bone resorption is also unclear. More in vitro and in vivo studies are recommended to fully elucidate direct and indirect actions of both molecules.

4. Materials and Methods

4.1. Study Population and Experimental Design

The study was designed as a single-center, randomized, prospective, controlled clinical trial. It was performed according to the Helsinki Declaration after previous acceptance obtained from the local bioethical committee (Bioethical Committee, Medical University of Bialystok, Poland R-I-002-397-2016). Each patient entering this research project signed an informed consent form. Thirty-nine generally healthy adults diagnosed with stage III periodontitis [33] were enrolled in the study. Thirty-eight (aged 24–73, mean age 45.4) were analyzed because of one patient's resignation. Among the analyzed participants, there were 21 women and 17 men.

The inclusion criteria were as follows:

- Presence of an intrabony defect, with a pocket depth (PD) $\geq$ 6 mm and a radiological defect depth of $\geq$3 mm and width of $\geq$2 mm;
- Over 18 years of age;
- Full mouth plaque index (FMPI) < 20% and full mouth bleeding on probing (FMBOP) < 20% [34,35].

Patients with general diseases that could affect the healing process, smokers, as well as pregnant or breastfeeding women were excluded.

Allocation of patients to test (M-MIST + Er:YAG + Nd:YAG) and control (M-MIST) groups was performed through a coin toss just before the surgery. Each participant had only one intrabony defect treated in the project. The patient and the surgeon were not blinded in the protocol because of technical reasons.

4.2. Clinical Examinations, Surgery, and Postoperative Care

The following clinical parameters were measured before surgery and 6 months post-op for each tooth with an intrabony defect: probing depth (PD), gingival recession (GR), and clinical attachment level (CAL). Each tooth was probed at six points (mesial, middle, and distal on both buccal and lingual sides). The cemento-enamel junction (CEJ) or the filling margin were taken as the reference points.

The full mouth plaque score (FMPS) and full mouth bleeding on probing (FMBOP) were calculated as a percentage based on the four surfaces of each tooth.

All measurements were performed with the use of a periodontal probe (PCP UNC15, Hu-Friedy, Chicago, IL, USA) and taken by the same experienced and calibrated examiner.

Intraoral radiographs were captured at baseline and after 6 months. To guarantee precision, a long cone parallel technique positioner was custom-prepared for each of the enrolled patients.

All surgical interventions were performed under local anesthesia (Septanest 100, Septodont, Paris, France). The surgical procedure consisted of intrasulcular incisions and preparation of the mucoperiosteal flap according to the principles of papilla preservation [56,57] and minimally invasive techniques [58].

In the test group, additionaly to modified minimally invasive surgery (M-MIST) dental laser (Fotona Light Walker AT-S, Dallas, TX, USA) was used.The granulation tissue was excised using the Er:YAG laser with parameters set at 3 W, 150 mJ, and 20 Hz LP, while the debridement of the root surface was accomplished with the same Er:YAG laser, operating at 1.6 W, 160 mJ, and 10 Hz LP. Concluding the surgical procedure, the formation of a blood clot was induced utilizing the Nd:YAG laser, configured at 2 W and 20 Hz VLP.

In the control group, the elimination of granulation tissue and the scaling and planing of root surfaces were carried out using manual instruments, specifically Gracey curettes (Hu-Friedy, Chicago, IL, USA) and ultrasonic scalers (EMS Piezon Tip PS, EMS, Nyon, Switzerland).

Once the debridement of the intrabony defect was completed, the mucoperiosteal flap was repositioned and stabilized by means of vertical modified mattress sutures (Ethilon 5.0, Johnson & Johnson Company, New Brunswick, NJ, USA).

Tooth mobility was an indication of splinting.

After surgery, patients were instructed to rinse the mouth twice daily using 0.2% chlorhexidine solution (Eludril, Pierre Fabre Laboratories, Paris, France), refrain from eating hard food, and avoid vigorous tooth brushing at the surgical area for 2 weeks. The sutures were removed 2 weeks post-op. Check-up appointments were scheduled for 1, 2, and 4 weeks and then at 3 and 6 months. Healing and possible complications (flap dehiscence, flap or papillae necrosis, suppuration, inflammation, as well as pain exacerbations) were monitored during the follow-up appointments. Additionally, at check-ups, the supragingival plaque was removed, and photographs of the surgical area were taken at every visit.

4.3. GCF Sampling

From the periodontal pocket corresponding to the operated intrabony defect, GCF was collected to determine the sulcus fluid flow rate (SFFR) in relative Periotron-units (PU) and to investigate IL-8 and MMP-9 levels. After the isolation of the tooth with cotton rolls and air-drying, the visible dental plaque was removed. Then, paper strips (Periopaper, Interstate Drug Exchange, Amityville, NY, USA) were placed in the periodontal pocket at a 1–2 mm depth for 30 s. The blood-contaminated strips were thrown away. The GCF volume absorbed on a paper strip (SFFR) was measured using a calibrated device (Periotron 8010, Oraflow, Plainview, NY, USA) and expressed in Periotron Units (PU). After measurement, the samples were immediately placed in Eppendorf tubes containing 200 µL of phosphate-buffered saline (PBS) and frozen.

4.4. GCF IL-8 and MMP-9 Analysis

The gingival crevicular fluid (GCF) samples collected at baseline and after 2 weeks, 4 weeks, and 6 months were used for laboratory analysis. The concentrations of IL-8 and MMP-9 in GCF were determined with the use of commercially available ELISA kits (Human CXCL8/IL8, R&D Systems, Minneapolis MN, USA and Human MMP-9 Elisa kit, R&D Systems, Minneapolis, MN, USA) following the manufacturer's instructions. The results were presented as the amount per 30 s per measurement point and expressed in pg/mL for IL-8 and ng/mL for MMP-9. We reported the total mediator content per 30 s sample because it removes a potential source of error from the analysis [59] and it is widely used in GCF elaborations [60].

4.5. Statistical Analysis

In the statistical analysis, the normality of distribution was verified using the Shapiro –Wilk test and the Kolgomorov–Smirnov tests with Lillefors correction. The Wilcoxon pair test was used to compare dependent variables over time. The nonparametric U Mann–Whitney test was used to compare the quantitative independent variables without distribution normality. ANOVA Friedman's test with Kendall's coefficient was used for multiple comparisons. Spearman's rank correlation coefficient was also determined.

The results were considered statistically significant for $p < 0.05$. Statistica 13.3 (TIBCO Software Inc. Palo Alto, CA, USA) was employed for calculations.

The sample size was calculated a priori by making an assumption about a standard deviation of a CAL change of 1 mm and to detect a mean difference of 1 mm with a test power of 80% on 32 subjects. However, considering possible drop-outs, 39 patients were recruited and randomized to the study.

5. Conclusions

Within the limits of our study, the results indicate that the additional use of Er:YAG + Nd:YAG lasers with the M-MIST procedure does not improve the clinical and biochemical treatment outcomes compared to M-MIST alone.

Author Contributions: Conceptualization, M.P.; methodology, M.P. and V.D.-P.; software, R.M.; validation, M.P., E.D., A.S., V.D.-P. and R.M.; formal analysis, R.M.; investigation, E.D., A.S. and V.D.-P.; data curation, E.D. and A.S.; writing—original draft preparation, E.D. and A.S.; writing—review and editing, M.P.; supervision, M.P.; project administration, E.D. and A.S.; funding acquisition, M.P. All authors have read and agreed to the published version of the manuscript.

Funding: The authors and their respective institution funded the study (Medical University of Bialystok, ul. Kilińskiego 1, 15-089 Bialystok, Poland—grant numbers N/ST/ZB/16/005/1164 and N/ST/ZB/17/002/1164). BTL Polska sp.z.o.o. rented the Fotona laser at no cost.

Institutional Review Board Statement: The study was compliant with the 1975 Helsinki Declaration and its 2000 amendments. The Ethical Committee of the Medical University of Bialystok approved this study (R-I-002-397-2016).

Informed Consent Statement: Each participant gave their informed consent prior to the study.

Data Availability Statement: The datasets used and/or analyzed during the current study are available from the corresponding author upon reasonable request.

Conflicts of Interest: The authors declare no conflicts of interest.

References

1. Chapple, I.L.C.; Mealey, B.L.; Van Dyke, T.E.; Bartold, P.M.; Dommisch, H.; Eickholz, P.; Geisinger, M.L.; Genco, R.J.; Glogauer, M.; Goldstein, M.; et al. Periodontal health and gingival diseases and conditions on an intact and a reduced periodontium: Consensus report of workgroup 1 of the 2017 World Workshop on the Classification of Periodontal and Peri-Implant Diseases and Conditions. *J. Periodontol.* **2018**, *89* (Suppl. S1), S74–S84. [CrossRef] [PubMed]
2. Billings, M.; Holtfreter, B.; Papapanou, P.N.; Mitnik, G.L.; Kocher, T.; Dye, B.A. Age-dependent distribution of periodontitis in two countries:findings from NHANES 2009 to 2014 and SHIP-TREND 2008 to 2012. *J. Clin. Periodontol.* **2018**, *45* (Suppl. S20), S130–S148. [CrossRef] [PubMed]
3. Fisher, J.; Selikowitz, H.S.; Mathur, M.; Varenne, B. Strengthening oral health for universal health coverage. *Lancet* **2018**, *392*, 899–901. [CrossRef] [PubMed]
4. Papapanou, P.N.; Tonetti, M.S. Diagnosis and epidemiology of periodontal osseous lesions. *Periodontology 2000* **2000**, *22*, 8–21. [CrossRef] [PubMed]
5. Kim, C.K.; Choi, S.H.; Kim, T.S.; Kaltschmitt, J.; Eickholz, P. The infrabony defect and its determinants. *J. Periodontal Res.* **2006**, *41*, 498–502. [CrossRef] [PubMed]

6. Papapanou, P.N.; Sanz, M.; Buduneli, N.; Dietrich, T.; Feres, M.; Fine, D.H.; Flemmig, T.F.; Garcia, R.; Giannobile, W.V.; Graziani, F.; et al. Periodontitis: Consensus report of workgroup 2 of the 2017 World Workshop on the Classification of Periodontal and Peri-Implant Diseases and Conditions. *J. Periodontol.* **2018**, *89* (Suppl. S1), S173–S182. [CrossRef]
7. Iorio-Siciliano, V.; Andreuccetti, G.; Blasi, A.; Matarasso, M.; Sculean, A.; Salvi, G.E. Clinical outcomes following regenerative therapy of non-contained intrabony defects using a deproteinized bovine bone mineral combined with either enamel matrix derivative or collagen membrane. *J Periodontol.* **2014**, *85*, 1342–1350. [CrossRef] [PubMed]
8. Iorio-Siciliano, V.; Blasi, A.; Nuzzolo, P.; Matarasso, M.; Isola, G.; Ramaglia, L. Treatment of periodontal intrabony defects using enamel matrix derivative: Surgical reentry after an observation period of at least 5 years. *Int. J. Periodontics Restor. Dent.* **2019**, *39*, 537–543. [CrossRef] [PubMed]
9. Stavropoulos, A.; Bertl, K.; Spineli, L.M.; Sculean, A.; Cortellini, P.; Tonetti, M. Medium- and long-term clinical benefits of periodontal regenerative/reconstructive procedures in intrabony defects: Systematic review and network meta-analysis of randomized controlled clinical studies. *J. Clin. Periodontol.* **2021**, *48*, 410–430. [CrossRef]
10. Thamaraiselvan, M.; Narayan, S.; Soundarajan, S. Minimally Invasive Surgery Periodontal Therapy for the Treatment of Intrabony Periodontal Defects: A Systematic Review. *Contemp. Clin. Dent.* **2022**, *13*, 101–107. [CrossRef]
11. Cortellini, P.; Tonetti, M.S. Improved wound stability with a modified minimally invasive surgical technique in the regenerative treatment of isolated interdental intrabony defects. *J. Clin. Periodontol.* **2009**, *36*, 157–163. [CrossRef]
12. Windisch, P.; Orban, K.; Salvi, G.E.; Sculean, A.; Molnar, B. Vertical-guided bone regeneration with a titanium-reinforced d-PTFE membrane utilizing a novel split-thickness flap design: A prospective case series. *Clin Oral Investig.* **2021**, *25*, 2969–2980. [CrossRef] [PubMed]
13. Ishikawa, I.; Aoki, A.; Takasaki, A.A.; Mizutani, K.; Sasaki, K.M.; Izumi, Y. Application of lasers in periodontics: True innovation or myth? *Periodontology 2000* **2009**, *50*, 90–126. [CrossRef] [PubMed]
14. Theodoro, L.H.; Marcantonio, R.A.C.; Wainwright, M.; Garcia, V.G. LASER in periodontal treatment: Is it an effective treatment or science fiction? *Braz. Oral Res.* **2021**, *35* (Suppl. S2), e099. [CrossRef] [PubMed]
15. Walsh, J.T., Jr.; Deutsch, T.F. Er:YAG laser ablation of tissue: Measurement of ablation rates. *Lasers Surg. Med.* **1989**, *9*, 327–337. [CrossRef]
16. Eberhard, J.; Ehlers, H.; Falk, W.; Açil, Y.; Albers, H.K.; Jepsen, S. Efficacy of subgingival calculus removal with Er:YAG laser compared to mechanical debridement: An in situ study. *J. Clin. Periodontol.* **2003**, *30*, 511–518. [CrossRef]
17. Aoki, A.; Ando, Y.; Watanabe, H.; Ishikawa, I. In vitro studies on laser scaling of subgingival calculus with an erbium:YAG laser. *J. Periodontol.* **1994**, *65*, 1097–1106. [CrossRef]
18. Crespi, R.; Romanos, G.E.; Cassinelli, C.; Gherlone, E. Effects of Er:YAG laser and ultrasonic treatment on fibroblast attachment to root surfaces: An in vitro study. *J. Periodontol.* **2006**, *77*, 1217–1222. [CrossRef]
19. Roncati, M.; Gariffo, A. Systematic review of the adjunctive use of diode and Nd:YAG lasers for nonsurgical periodontal instrumentation. *Photomed. Laser Surg.* **2014**, *32*, 186–197. [CrossRef] [PubMed]
20. Zhu, J.; Wei, R.; Lv, X.; Qu, C. Efficacy of a combined Er:YAG laser and Nd:YAG laser in non-surgical treatment for severe periodontitis. *Lasers Med. Sci.* **2022**, *37*, 1095–1100. [CrossRef] [PubMed]
21. Estrin, N.E.; Moraschini, V.; Zhang, Y.; Romanos, G.E.; Sculean, A.; Miron, R.J. Combination of Nd:YAG and Er:YAG lasers in non-surgical periodontal therapy: A systematic review of randomized clinical studies. *Lasers Med. Sci.* **2022**, *37*, 2737–2743. [CrossRef] [PubMed]
22. Van Dyke, T.E. Shifting the paradigm from inhibitors of inflammation to resolvers of inflammation in periodontitis. *J. Periodontol.* **2020**, *91* (Suppl. S1), S19–S25. [CrossRef] [PubMed]
23. Golub, L.M.; Lee, H.M. Periodontal therapeutics: Current host-modulation agents and future directions. *Periodontology 2000* **2020**, *82*, 186–204. [CrossRef] [PubMed]
24. Baggiolini, M.; Walz, A.; Kunkel, S.L. Neutrophil-activating peptide-1/interleukin 8, a novel cytokine that activates neutrophils. *J. Clin. Investig.* **1989**, *84*, 1045–1049. [CrossRef] [PubMed]
25. Tonetti, M.S. Molecular factors associated with compartmentalization of gingival immune responses and transepithelial neutrophil migration. *J. Periodontal Res.* **1997**, *32*, 104–109. [CrossRef] [PubMed]
26. Sorsa, T.; Tjäderhane, L.; Salo, T. Matrix metalloproteinases (MMPs) in oral diseases. *Oral Dis.* **2004**, *10*, 311–318. [CrossRef] [PubMed]
27. Elgezawi, M.; Haridy, R.; Almas, K.; Abdalla, M.A.; Omar, O.; Abuohashish, H.; Elembaby, A.; Wölfle, C.U.; Siddiqui, Y.; Kaisarly, D. Matrix metalloproteinases in dental and periodontal tissues and their current inhibitors: Developmental, degradational and pathological aspects. *Int. J. Mol. Sci.* **2022**, *23*, 8929. [CrossRef] [PubMed]
28. Uitto, V.J.; Overall, C.M.; McCulloch, C. Proteolytic host cell enzymes in gingival crevice fluid. *Periodontology 2000* **2003**, *31*, 77–104. [CrossRef] [PubMed]
29. Pozo, P.; Valenzuela, M.A.; Melej, C.; Zaldivar, M.; Puente, J.; Martinez, B.; Gamonal, J. Longitudinal analysis of metalloproteinases, tissue inhibitors of metalloproteinases and clinical parameters in gingival crevicular fluid from periodontitis-affected patients. *J. Periodontal Res.* **2005**, *40*, 199–207. [CrossRef] [PubMed]
30. Marcaccini, A.M.; Meschiari, C.A.; Zuardi, L.R.; de Sousa, T.S.; Taba, M.; Teofilo, J.M.; Jacob-Ferreira, A.L.B.; Tanus-Santos, J.E.; Novaes, A.B.; Gerlach, R.F. Gingival crevicular fluid levels of MMP-8, MMP-9, TIMP-2, and MPO decrease after periodontal therapy. *J. Clin.Periodontol.* **2010**, *37*, 180–190. [CrossRef] [PubMed]

31. Hill, P.A.; Docherty, A.J.; Bottomley, K.M.; O'Connell, J.P.; Morphy, J.R.; Reynolds, J.J.; Meikle, M.C. Inhibition of bone resorption in vitro by selective inhibitors of gelatinase and collagenase. *Biochem. J.* **1995**, *308*, 167–175. [CrossRef] [PubMed]
32. Hernández Ríos, M.; Sorsa, T.; Obregón, F.; Tervahartiala, T.; Valenzuela, M.A.; Pozo, P.; Dutzan, N.; Lesaffre, E.; Molas, M.; Gamonal, J. Proteolytic roles of matrix metalloproteinase (MMP)-13 during progression of chronic periodontitis: Initial evidence for MMP-13/MMP-9 activation cascade. *J. Clin. Periodontol.* **2009**, *36*, 1011–1017. [CrossRef] [PubMed]
33. Tonetti, M.S.; Greenwell, H.; Kornman, K.S. Staging and grading of periodontitis: Framework and proposal of a new classification and case definition. *J. Clin. Periodontol.* **2018**, *45* (Suppl. S20), S149–S161. [CrossRef] [PubMed]
34. O'Leary, T.J.; Drake, R.B.; Naylor, J.E. The plaque control record. *J. Periodontol.* **1972**, *43*, 38. [CrossRef] [PubMed]
35. Ainamo, J.; Bay, I. Problems and proposals for recording gingivitis and plaque. *Int. Dent. J.* **1975**, *25*, 229–235. [PubMed]
36. Patel, R.A.; Wilson, R.F.; Palmer, R.M. The effect of smoking on periodontal bone regeneration: A systematic review and meta-analysis. *J. Periodontol.* **2012**, *83*, 143–155. [CrossRef] [PubMed]
37. Reynolds, M.A.; Kao, R.T.; Camargo, P.M.; Caton, J.G.; Clem, D.S.; Fiorellini, J.P.; Geisinger, M.L.; Mills, M.P.; Nares, S.; Nevins, M.L. Periodontal regeneration—Intrabony defects: A consensus report from the AAP Regeneration Workshop. *J. Periodontol.* **2015**, *86*, S105–S107. [CrossRef]
38. Mizutani, K.; Aoki, A.; Coluzzi, D.; Yukna, R.; Wang, C.Y.; Pavlic, V.; Izumi, Y. Lasers in minimally invasive periodontal and peri-implant therapy. *Periodontology 2000* **2016**, *71*, 185–212. [CrossRef] [PubMed]
39. Koidou, V.P.; Chatzopoulos, G.S.; Tomas, I.; Nibali, L.; Donos, N. Expression of gingival crevicular fluid markers during early and late healing of intrabony defects after surgical treatment: A systematic review. *Clin. Oral. Investig.* **2020**, *24*, 487–502. [CrossRef] [PubMed]
40. Barros, S.P.; Williams, R.; Offenbacher, S.; Morelli, T. Gingival crevicular fluid as a source of biomarkers for periodontitis. *Periodontology 2000* **2016**, *70*, 53–64. [CrossRef]
41. Griffiths, G. Formation, collection and significance of gingival crevice fluid. *Periodontology 2000* **2003**, *31*, 32–42. [CrossRef] [PubMed]
42. Armitage, G.C. Analysis of gingival crevicular fluid and risk of progression of Periodontitis. *Periodontology 2000* **2004**, *34*, 109–119. [CrossRef] [PubMed]
43. Kuru, L.; Griffiths, G.S.; Petrie, A.; Olsen, I. Changes in transforming growth factor-beta1 in gingival crevicular fluid following periodontal surgery. *J. Clin. Periodontol.* **2004**, *31*, 527–533. [CrossRef] [PubMed]
44. Okuda, K.; Miyazaki, A.; Momose, M.; Murata, M.; Nomura, T.; Kubota, T.; Wolff, L.F.; Yoshie, H. Levels of tissue inhibitor of metalloproteinases-1 and matrix metalloproteinases-1 and -8 in gingival crevicular fluid following treatment with enamel matrix derivative(Emdogain). *J. Periodontal Res.* **2001**, *36*, 309–316. [CrossRef] [PubMed]
45. Dolińska, E.; Pietruska, M.; Dymicka-Piekarska, V.; Milewski, R.; Sculean, A. Matrix metalloproteinase 9 (MMP-9) and interleukin-8 (IL-8) in gingival crevicular fluid after regenerative therapy in periodontal intrabony defects with and without systemic antibiotics-randomized clinical trial. *Pathogens* **2022**, *11*, 1184. [CrossRef] [PubMed]
46. Koidou, V.P.; Hagi-Pavli, E.; Cross, S.; Nibali, L.; Donos, N. Molecular profiling of intrabony defects' gingival crevicular fluid. *J. Periodontal Res.* **2022**, *57*, 152–161. [CrossRef] [PubMed]
47. Koidou, V.P.; Hagi-Pavli, E.; Nibali, L.; Donos, N. Elucidating the molecular healing of intrabony defects following non-surgical periodontal therapy: A pilot study. *J. Periodontal Res.* **2023**, *59*, 53–62. [CrossRef] [PubMed]
48. Pellegrini, G.; Rasperini, G.; Pagni, G.; Giannobile, W.V.; Milani, S.; Musto, F.; Dellavia, C. Local wound healing biomarkers for real-time assessment of periodontal regeneration: Pilot study. *J. Periodontal Res.* **2017**, *52*, 388–396. [CrossRef] [PubMed]
49. Rakmanee, T.; Calciolari, E.; Olsen, I.; Darbar, U.; Griffiths, G.S.; Petrie, A.; Donos, N. Expression of growth mediators in the gingival crevicular fluid of patients with aggressive periodontitis undergoing periodontal surgery. *Clin. Oral. Investig.* **2018**, *23*, 3307–3318. [CrossRef] [PubMed]
50. Mäkelä, M.; Salo, T.; Uitto, V.J.; Larjava, H. Matrix metalloproteinases (MMP-2 and MMP-9) of the oral cavity: Cellular origin and relationship to periodontal status. *J. Dent. Res.* **1994**, *73*, 1397–1406. [CrossRef] [PubMed]
51. Franco, C.; Patricia, H.R.; Timo, S.; Claudia, B.; Marcela, H. Matrix metalloproteinases as regulators of periodontal inflammation. *Int. J. Mol. Sci.* **2017**, *18*, 440. [CrossRef] [PubMed]
52. Hernández, M.; Dutzan, N.; García-Sesnich, J.; Abusleme, L.; Dezerega, A.; Silva, N.; González, F.E.; Vernal, R.; Sorsa, T.; Gamonal, J. Host-pathogen interactions in progressive chronic periodontitis. *J. Dent. Res.* **2011**, *90*, 1164–1170. [CrossRef] [PubMed]
53. Hernandez, M.; Valenzuela, M.A.; Lopez-Otin, C.; Alvarez, J.; Lopez, J.M.; Vernal, R.; Gamonal, J. Matrix metalloproteinase-13 is highly expressed in destructive periodontal disease activity. *J. Periodontol.* **2006**, *77*, 1863–1870. [CrossRef] [PubMed]
54. Sijari, Z.; Romano, F.; Ciardo, G.; Martella, V.; Maggiora, M.; Bottone, M.; Aimetti, M. Effect of two post-surgical cleansing protocols on early periodontal wound healing and cytokine levels following osseous resective surgery: A randomized controlled study. *Int. J. Dent. Hyg.* **2019**, *17*, 300–308. [CrossRef] [PubMed]
55. Opdenakker, G.; Van den Steen, P.E.; Dubois, B.; Nelissen, I.; Van Coillie, E.; Masure, S.; Proost, P.; Van Damme, J. Gelatinase B functions as regulator and effector in leukocyte biology. *J. Leukoc. Biol.* **2001**, *69*, 851–859. [CrossRef] [PubMed]
56. Cortellini, P.; Pini-Prato, G.; Tonetti, M.S. The modified papilla preservation technique. A new surgical approach for interproximal regenerative procedures. *J. Periodontol.* **1995**, *66*, 261–266. [CrossRef] [PubMed]
57. Cortellini, P.; Pini-Prato, G.; Tonetti, M.S. The simplified papilla preservation flap. A novel surgical approach for the management of soft tissues in regenerative procedures. *Int. J. Periodont. Rest. Dent.* **1999**, *19*, 589–599.

58. Cortellini, P.; Tonetti, M.S. Microsurgical approach to periodontal regeneration. Initial evaluation in a case cohort. *J. Periodontol.* **2001**, *72*, 559–569. [CrossRef] [PubMed]
59. Wassall, R.R.; Preshaw, P.M. Clinical and technical considerations in the analysis of gingival crevicular fluid. *Periodontology 2000* **2016**, *70*, 65–79. [CrossRef] [PubMed]
60. Lamster, I.B.; Ahlo, J.K. Analysis of gingival crevicular fluid as applied to the diagnosis of oral and systemic diseases. *Ann. N. Y. Acad. Sci.* **2007**, *1098*, 216–229. [CrossRef] [PubMed]

antibiotics

MDPI

Article

Prevalence and Phylogenetic Analysis of Lipoprotein-Gene *ragB-1* of *Porphyromonas gingivalis*—A Pilot Study

Sarah Böcher [1,*], Hendrik L. Meyer [2], Evdokia Dafni [2] and Georg Conrads [2]

[1] Department of Operative Dentistry, Periodontology and Preventive Dentistry, Rheinisch-Westfälische Technische Hochschule (RWTH) University Hospital, Pauwelsstrasse 30, 52074 Aachen, Germany

[2] Division of Oral Microbiology and Immunology, Department of Operative Dentistry, Periodontology and Preventive Dentistry, Rheinisch-Westfälische Technische Hochschule (RWTH) University Hospital, Pauwelsstrasse 30, 52074 Aachen, Germany; gconrads@ukaachen.de (G.C.)

* Correspondence: sboecher@ukaachen.de; Tel.: +49-241-80-88110

Abstract: *Porphyromonas gingivalis* (*P.g.*) is a key pathogen involved in periodontal diseases. The aim of this study was to investigate the prevalence and phylogenetic origin of the lipoprotein-gene *ragB* in its most virulent variant, *ragB-1* (co-transcribed with *ragA-1* as locus *rag-1*), in different *P.g.* strains collected worldwide. A total of 138 *P.g.* strains were analyzed for the presence of *ragB-1* by pooled analysis and subsequently individual PCRs. Sequencing a core fragment of *ragB-1* of the individual strains made it possible to carry out a phylogenetic classification using sequence alignment. In total, 22 of the 138 *P.g.* strains tested positive for *ragB-1*, corresponding to a prevalence of 16%. The fragment investigated was highly conserved, with variations in the base sequence detected in only three strains (OMI 1072, OMI 1081, and OMI 1074). In two strains, namely OMI 1072 (original name: I-433) and OMI 1081 (original name: I-372), which originate from monkeys, two amino-acid alterations were apparent. Since *ragB-1* has also been found in animal strains, it may be concluded that *rag-1* was transferred from animals to humans and that this originally virulent variant was weakened by mutations over time so that new, less virulent, adapted commensal versions of *rag* (*rag-2*, *-3*, *and -4*), with *P.g.* as the host, evolved.

Keywords: *Porphyromonas gingivalis*; phylogeny; lipoprotein RagB; periodontal diseases; virulence

1. Introduction

Porphyromonas gingivalis (*P. gingivalis*, *P.g.*) is a black-pigmented, immobile, gram-negative, anaerobic rod [1], which is considered a key pathogen (or pathobiont) involved in periodontal diseases. There is considerable heterogeneity between the different isolates, including variable virulence [2]. Molecular typing studies have shown that strains of certain genotypes are more frequently associated with disease (periodontitis) [3–7]. Animal models have also confirmed that some strains are more pathogenic than others. However, definitive factors specific to virulent or avirulent *P.g.* strains have not yet been isolated. Well-known virulence factors do not fully explain the differences in pathogenicity between different strains [2,7,8]. Deeper genetic analysis of *P.g.* is therefore crucial for understanding its properties and the role of its individual virulence factors [9]. A better understanding of the factors that determine the variance of strains, in terms of their pathogenic potential, is, in turn, important for improving diagnostic tools and therapeutic strategies [7].

RagA and RagB proteins are major components of the outer membrane of *P.g.* and have been associated with *P.g.* virulence by contributing to subcutaneous lesion development, epithelial cell invasion, and efficient growth of the pathogen by acting as a transport system for nutrients [10] but have been little studied in comparison with other virulence factors. They are proteins that are 115 (RagA) and 55 kDa (RagB) in size, encoded by genes PG0185 (*ragA*) (3.1 kb) and PG0186 (*ragB*) (1.5 kb) [11–15]. The *ragA* and *ragB* genes are co-transcribed to a single, approximately 4.7 kb, mRNA [16], with *ragB* located immediately

(30 bp) downstream of the *ragA* gene [7]. Mutants with deletion of either *ragA* or *ragB* have been shown to be phenotypically negative for both proteins [17]. RagAB occurs as a receptor pair on the bacterial cell surface of *P.g.* and consists of the TonB-dependent-transporter (TBDT) RagA (PG0185) and the surface-lipoprotein RagB (PG0186) [10,18]. The protein pair is also considered to exhibit typical features of a TonB-dependent outer-membrane receptor [13,14].

The X-ray crystal structure of RagAB, purified from *P.g.* W83, revealed a dimeric $RagA_2B_2$ complex with an architecture similar to that of SusCD (starch-utilization system), a TBDT carbohydrate transporter in the outer membrane of *Bacteroides thetaiotaomicron* (*B. thetaiotaomicron*) [7,10,18,19], with RagA showing substantial homology to SusC and RagB to SusD. In this heterotetrameric complex, the single RagA molecule has the structure of a 22-stranded β-barrel with an inserted plug domain, similar to other TBDTs, and is tightly capped on the extracellular side by RagB, which covers a large surface area (approximately 3,850 $Å^2$). Full-length RagB contains a signal peptide for export to the outer membrane and a cysteine that is lipidated and anchors the protein to the cell membrane. RagB is a compact molecule of $70 \times 55 \times 50$ Å in size and consists of a single domain centered on a curved helical scaffold. This scaffold consists of ten large α-helices, of which the first eight form four tetratricopeptide repeats (TPRs), each arranged as two helices connected by a linker. Thus, RagB is a tetratricopeptide-repeat protein (TPRP). The TPRs form a right-handed solenoid of helices that is curved and complemented by two capping helices located downstream. The concave surface also bears four large intertwined irregular inserts (A–D) [10,18,20].

RagAB together form a large, closed internal cavity, containing a molecule of approximately 13 residues long bound to the RagAB interface, which can be considered direct evidence for a "pedal bin" mechanism [18]. The binding of an extracellular substance (peptide) leads to a conformational change that causes disruption of the N-terminal Ton box on the periplasmatic side of the plug domain, making it accessible for interaction with TonB (an inner membrane protein complex). The resulting disruption of the plug domain, in turn, allows the formation of a transport channel to the periplasmatic space that enables translocation of the substrate (Figure 1). Hence, the binding of a peptide to RagAB causes the lid to close and the complex to enter into a transport state. The disrupted state of the Ton box can therefore be considered a signal that ensures that only substrate-loaded transporters form productive complexes with TonB [18].

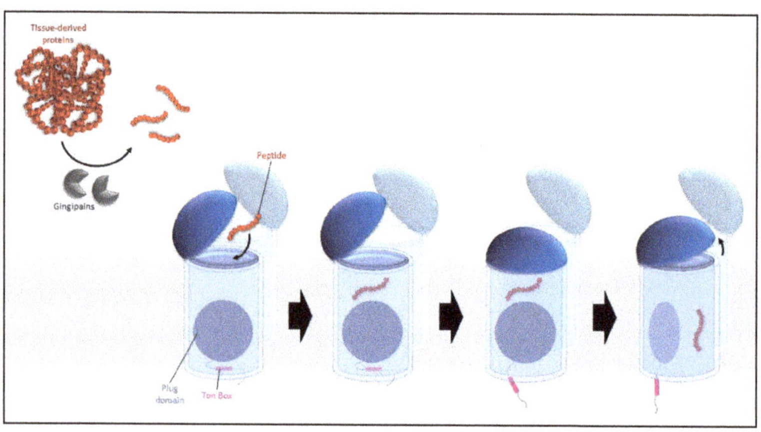

Figure 1. Schematic illustration of the function of RagAB (modified after [18]).

Substantial structural similarities of RagB with the *B. thetaiotaomicron* SusD and *Tannerella forsythia* (*T. forsythia*, T.f.) NanU (neuraminate uptake system) have already been confirmed. Since the crystal structure of RagB that has been determined shows putative saccharide-binding sites on the molecular surface, as well as bound monosaccharides [10,18,20], it was originally assumed that RagB (together with RagA) is involved in sugar binding and uptake. Although *P.g.* is an asaccharolytic bacterium, this does not mean that it does not require sugar intake. The uptake of saccharides is also necessary for reasons other than energy acquisition—for example, for capsule formation or self-glycosylation of proteins [10,14]. However, analysis of the combined RagAB complex indicated that RagAB is actually a dynamic oligopeptide "acquisition machine" located at the outer membrane, with considerable substrate selectivity, essential for peptide uptake by *P.g.* and effective consumption of proteinaceous nutrients [10,18]. Nagano et al. [12] constructed mutants lacking one—or both—of these two genes (*ragA, ragB,* or *ragAB*) and showed that double-deletion mutants lacked both RagA and RagB, while mutants lacking *ragA* showed significantly reduced RagB expression and mutants lacking *ragB* produced only degraded RagA. All mutants grew normally in a nutrient-rich medium and medium containing already digested protein. In a medium containing undigested native protein, growth of the mutants lacking *ragA* and *ragAB* was significantly slower because *P.g.* had to metabolize them first to meet nutritional requirements; however, the amounts and activities of gingipains were unaltered. The authors therefore assumed that mutants lacking RagA and RagB may not be able to take up larger peptides and thus need more time to grow in a medium containing native protein, since smaller peptides must first become available through digestion by the proteases. This confirmed the assumption that cell-surface-associated RagA and RagB are involved in the transport of macromolecules, such as protein-degradation products (peptides), so that strains/mutants lacking these surface molecules grow more slowly due to the lower substrate supply [7,11,21]. This transport is selective, as only certain peptides bind to RagAB and are transported through the outer membrane [18].

RagA, and RagB in particular, are immunodominant antigens [21–23]. Early studies have already shown that the IgG response to RagB (determined by serum antibody detection) was higher in adults with periodontitis than in healthy controls [10,21]. More recent studies have also found a strong systemic RagB-recognizing antibody response in individuals with periodontitis [10,22,23]. In vitro studies exposing human inflammatory cells to purified RagB revealed that the expression of several genes encoding pro-inflammatory mediators in monocytes (such as IL-1α, IL-1β, IL-6, and IL-8) was induced in a dose-dependent manner. RagB thus appears to act as a pro-inflammatory mediator. RagB mutants, on the other hand, appear to have a lower inflammatory capacity in comparison with wild-type *P.g.* W83 [10,24]. RagB may therefore play an important role in the etiology of *P.g.*-associated periodontal inflammation [10]. Indeed, expression profiles of RagAB appear to be related to periodontal pocket depth, with increasing expression in deeper pockets [10,14,16], and there seems to be a clear correlation between *rag* gene transcripts in gingival crevicular fluid and clinical indices of periodontitis [10,25].

Sequence analysis of the *ragAB* genes and flanking regions led to the conclusion that the genetic locus was acquired by horizontal gene transfer and that the genetic characteristics of the locus correspond to a pathogenicity island [14]. Thus, the *rag* locus has a low G+C content of 41% in comparison with the G+C content of the whole genome of *P.g.*, which, with its 2,343,479 bp, has an average G+C content of 48.3% [26]. In addition, the locus is flanked by insertion sequences, elements that may have influenced the original acquisition, also indicating a different "foreign" species as the source [14]. The presence of these atypical islands raises the question of how they got there and which unknown microorganism is the "foreign" source of these genes. Studies have shown that more than 40% of the protein sequences encoded in these regions show high homology to proteins of *B. thetaiotaomicron*, an enteric commensal [27,28]. Furthermore, it is conceivable that a gram-negative oral anaerobe acted as a mediator in the transmission [28]. In particular, the close and constant spatial relationships in dental plaque may provide favorable conditions for the transfer of conjugative transposons through cell-to-cell contact [28–30]. Thus, the DNA uptake ability has been shown to increase when bacteria are arranged in plaque-like biofilms [28,31–33].

However, further research is needed to fully understand the interactions between RagAB and the host immune response [10]. Additionally, a number of studies suggest that the *rag* locus may be a suitable therapeutic target for *P.g.*-associated diseases. Since RagB dominates the *P.g.* antibody response in humans and is an important virulence factor, it has been considered an attractive potential vaccine target. Interestingly, besides RagB, other TonB-dependent outer-membrane receptors have also been investigated as potential vaccine targets for a variety of bacteria [10,34–36]. The aim of this study was to determine the prevalence of the virulence-gene PG0186 (lipoprotein-gene *ragB* in its most virulent variant, *ragB-1*) in diverse *Porphyromonas gingivalis* strains collected worldwide. In addition, the exact sequence of a section of the *ragB-1* gene of the different strains was evaluated and a phylogenetic classification made by means of sequence alignment.

2. Results

In total, 22 of the 138 *P.g.* strains tested positive, which corresponds to a prevalence of 16%. An overview of all *ragB-1*-positive strains, providing the OMI and other strain numbers used before, as well as their origin (host, year of isolation, and country), is shown in Table 1. The *ragB-1* sequence from OMI 629 (W83) precisely matched the previously published sequence of W83 [37]. However, we found that the primers used did not cover the complete *ragB-1* gene but instead only a section/fragment of 436 bp long (Figure 2), as the complete *ragB-1* gene comprises 1056 bp.

Table 1. Overview of all *P. gingivalis*/*P. gulae* strains (22 out of 138, 16%) that tested positive for the *ragB-1* gene.

Strain (OMI)	Species	Original Code	Origin Species	Isolation Year	Country
629	*P. gingivalis*	W83	Human	1991	Bonn, Germany
1049	*P. gingivalis*	AJW5 (VAG 5)	Human	1991	Buffalo, NY, USA
1051	*P. gingivalis*	22KN6-12	Human		Tokushima, Japan
1060	*P. gingivalis/gulae*	OMG 1426	Monkey	1989	Florida, USA
1053	*P. gingivalis*	RB22D-1	Human		Quebec, Canada
1068	*P. gingivalis*	213Pg2	Human	1994	Indonesia
1071	*P. gingivalis*	ATCC49417, RB22D	Human	1993	Quebec, Canada
1072	*P. gingivalis/gulae*	I-433	Monkey	1989	Florida, USA
1074	*P. gingivalis*	7B5	Human		Quebec, Canada
1078	*P. gingivalis*	84Pg1-a	Human	1994	Indonesia
1079	*P. gingivalis*	83Pg1-a	Human	1994	Indonesia
1081	*P. gingivalis/gulae*	I-372	Monkey	1989	Florida, USA
1084	*P. gingivalis*	81Pg1-a	Human	1994	Indonesia
1087	*P. gingivalis*	122Pg1-a	Human	1994	Indonesia
1101	*P. gingivalis*	83Pg1-b	Human	2002	Indonesia
1108	*P. gingivalis*	81Pg1-b	Human	2002	Indonesia
1112	*P. gingivalis*	A 7436	Human		Georgia, USA
1117	*P. gingivalis*	W12	Human		Alabama, USA
1120	*P. gingivalis*	13JC	Human		Rennes, France
1122	*P. gingivalis*	17-5	Human		Minneapolis, MN, USA
1125	*P. gingivalis*	122Pg1-b	Human	2002	Indonesia
1127	*P. gingivalis*	84Pg1-b	Human	2002	Indonesia

Four *ragB-1* strains are persisting in patients (81, 83, 84, 122) from 1994 (a) to 2002 (b).

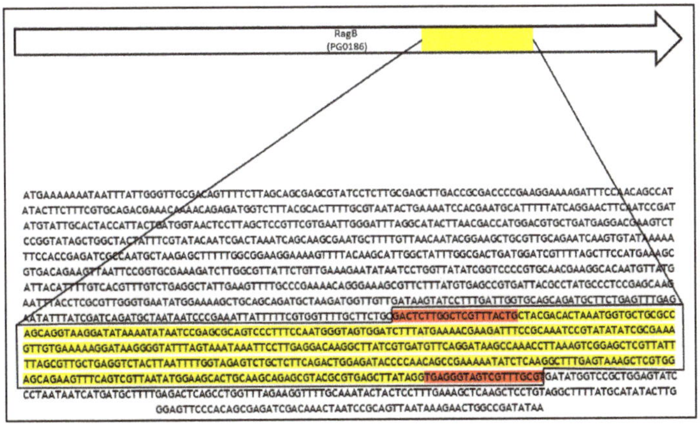

Figure 2. Nucleotide sequence of the *ragB-1* gene of *P. gingivalis* W83 [37]. The part of the *ragB-1* gene sequenced in the present study is highlighted in yellow, and the primer sequences are highlighted in red.

The sequence of the DNA segment examined in this study proved to be highly conserved. Overall, variations (point mutations, PMs) within the base sequence were only detected in three strains: OMI 1072, OMI 1081 (both six PMs), and OMI 1074 (a single PM). Figure 3 shows the individual base variations with the respective positions of the corresponding chromatograms. OMI 1072 and OMI 1081 showed the same six base variations and at exactly the same positions, while OMI 1074 showed only a single altered base at a completely different position. Strain OMI 629 (W83) served as a reference at the top.

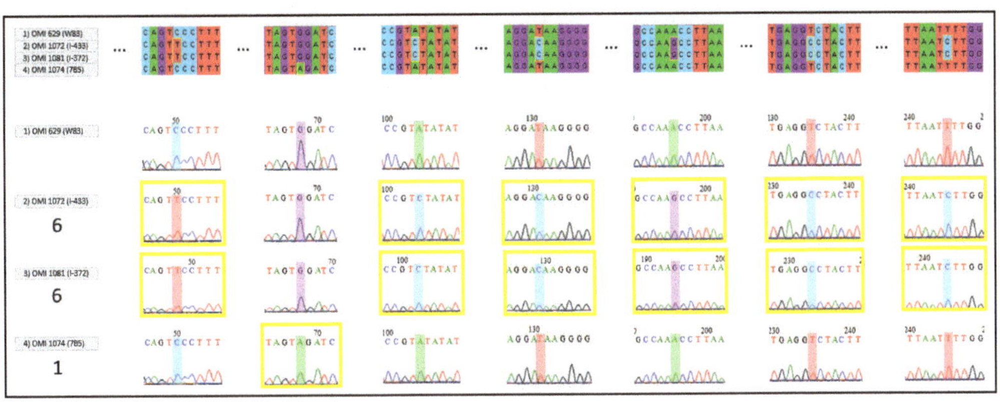

Figure 3. Shown are *ragB-1* variations (outlined in yellow) limited to three *P. gingivalis* strains (OMI 1072, OMI 1081, and OMI 1074). OMI 1072 and OMI 1081 showed the same six base variations at exactly the same positions, while OMI 1074 showed only a single altered base, at a completely different position. Strain OMI 629 (W83) served as a reference.

With regard to the translated amino-acid sequence, four out of six of the PMs detected in OMI 1072 and OMI 1081, as well as the one in OMI 1074, were found to be synonymous mutations, thus with no change in the resulting amino-acid sequence. The remaining two PMs detected in OMI 1072 and OMI 1081 lead, however, to the following changes in the amino-acid sequence: alanine was changed to valine (both nonpolar/hydrophobic amino acids) and serine changed to asparagine (both polar/neutral amino acids). The gene regions affected by the point mutations and the resulting amino-acid sequence are shown in Figure 4.

Figure 4. Amino-acid sequences corresponding to alterations found in *ragB-1*. Four of the point mutations detected in OMI 1072 and OMI 1081, as well as the one in OMI 1074, did not cause any change in the resulting amino-acid sequence (synonymous mutations, framed in red). Only two of the point mutations in OMI 1072 and OMI 1081 lead to changes in the translated amino-acid sequence (framed in yellow). Here, alanine was translated instead of valine (both nonpolar/hydrophobic amino acids) and serine was translated instead of asparagine (both polar/neutral amino acids), all conservative (neutral) mutations with no or little effect on RagB functionality.

The neighbor-joining (NJ) phylogenetic tree (Figure 5) shows a close relationship among all strains that tested positive for *ragB-1*, except for OMI 1072 (original name: I-433) and OMI 1081 (original name: I-372). These two strains originate from monkeys and were originally classified as *P.g.* However, in 2001, they were re-classified by Fournier et al. [38] as *P. gulae*, which is an animal *P. gingivalis*-like biotype. Interestingly, unlike human *P.g.* strains, these two *P. gulae* strains are catalase-positive [38].

Figure 5. The NJ phylogenetic tree calculated on the basis of 436 bp of *ragB-1* shows identity between all strains, except for OMI 1074 (7B5), OMI 1072 (I-433), and OMI 1081 (I-372), the latter two being *P. gulae* strains from monkeys (program Mega 11, version 11.0.13).

As an expected result of a standard protein BLAST (blastp), the entire sequence of the RagB-1 of *P.g.* W83 showed an analogy to the RagB-/SusD-family nutrient-uptake outer-membrane proteins. It particularly featured a significant analogy to the SusD of *B. thetaiotaomicron*, a TonB-dependent receptor.

Comparison with previously published 3D structural data for RagB-1 [20] shows that the part of RagB-1 investigated in this study (positions 319 to 413) represents part of one of the protein inserts (insert C) and α-helices nos. 12 to 15 [37].

3. Materials and Methods

A total of 138 *P.g.* strains available in the strain collection and genome database of the Division of Oral Microbiology and Immunology at RWTH Aachen University Hospital were analyzed for the presence of the lipoprotein-gene *ragB-1* by pooled analysis and later single PCRs, and the results were visualized by gel electrophoresis. Since only a small number of *ragB-1*-positive strains was expected, the 138 *P.g.* strains were first examined in a pooled test procedure, for which the DNA of 10 strains each was pooled. After these pooled PCRs, the strains of each pool that tested positive for *ragB-1* were individually retested for the presence of the *ragB-1* locus. PCRs were performed using a PG0186 primer set (PG0186-F 5′ GACTCTTGCTCGTTTACTG 3′ and PG0186-R 5′ ACGCAAACGACTAC-CCTCA 3′ [8], synthesized by TIB MOLBIOL Syntheselabor GmbH, Berlin, Germany). After all strains containing a *ragB-1* locus had been identified, the PCR amplicons were purified (NucleoSpin® Gel and PCR Clean-up set, Macherey-Nagel GmbH & Co. KG, Düren, Germany) and sent for DNA sequencing to a service laboratory (LightRun Tube Sequencing Service, Eurofins Genomics Sequencing GmbH, Cologne, Germany). Amplicons were sequenced using the forward primer or both forward and reverse primers and were then assembled into contigs using the Sequence Alignment Editor (BioEdit Sequence Alignment Editor version 7.2.5, Ibis Biosciences, Carlsbad, CA, USA [39,40]). Subsequently, sequence alignment was performed using the MEGA11 program (MEGA11: Molecular Evolutionary Genetics Analysis, version 11 [41]), and a phylogenetic tree ("Construct/Test Neighbor-Joining Tree") was created from the individual base sequences of all strains that tested positive for the *ragB-1* gene. To evaluate sequence homologies with other known proteins, the standard protein BLAST (blastp) of the National Center for Biotechnology Information (NCBI) was used [37].

4. Discussion

In the present study, 22 of 138 *P.g.* strains tested were positive for the presence of *ragB-1*, corresponding to a prevalence of 16%. This is consistent with previous findings that the *rag* locus has four variants and that the most virulent variant, *ragB-1*, is present in only a proportion of strains and clinical isolates [14]. Frandsen et al. [8] reported that the *ragB(-1)* locus was found in 13% of strains examined. This percentage is lower than the prevalence of 26% reported for *ragAB-1* in a study by Hall et al. among 168 isolates [7].

However, studies on the prevalence of the *ragB* gene should be interpreted with caution. For example, in the present study, only part of the *ragB(-1)* gene (436 bp) could be examined, as mentioned in the original publication [8]. In turn, different studies have described other primers that were used for the investigation of *ragB* and also examined only part of the *ragB(-1)* gene—which differed, however, from the part examined in the present study. For example, studies by Dolgilevich et al. and Bunte et al. [42,43] used primer pairs that allowed amplification of an upstream part of the *ragB(-1)* gene that was 432 bp long. In fact, the actual *ragB(-1)* gene is 1056 bp long in total, and only a part is covered by established primer pairs [37].

Because prevalence has been investigated on the basis of amplification of different parts of the *ragB* gene, the validity of the results of existing studies on the actual prevalence of the *ragB* gene is limited. Earlier analyses suggest that only a subset of *P.g.* strains contain the *rag* locus [8,14,44], but more-recent whole-genome sequencing clearly shows that the *ragAB* operon is present in probably all strains and clinical isolates analyzed to date. In

2005, Hall et al. [7] demonstrated that there are actually four different variants or alleles of the *rag* locus (*rag-1* to *rag-4*) (Figure 6), with *rag-1* representing the most virulent variant found in W50 and W83, *rag-2* representing the locus with strain A011/9 as a reference, *rag-3* representing the locus of strain QM220 as a reference, and *rag-4* representing the locus of type strain ATCC 33277. This is particularly important, as *rag-2*, *rag-3*, and *rag-4* have frequently been detected in isolates lacking *rag-1* [7,43]. Each of these alleles appears to be common in the population of *P.g.* and is not restricted to a specific geographical region [7]. In a study of 23 *P.g.* strains investigated, *rag-2* and *rag-4* were the most prevalent alleles [45], while Hall et al. [7] found that *rag-1* was the second most common allele among 168 clinical isolates. In total, 26% of these were carriers of *rag-1*, 36% of *rag-2*, 25% of *rag-3*, and 14% of *rag-4*. The presence of different alleles of the *rag* locus also leads to the question of whether certain alleles are associated with different clinical manifestations. In particular, *rag-1* appears to be associated with deep periodontal pockets. Strains shown to be more virulent in mouse models have also been shown to more likely carry *rag-1* than other alleles [7,46]. In another study, *rag-3* and *rag-4* were the predominant genotypes in patients with orthodontic gingivitis and mild-to-moderate forms of periodontitis [25]. The different variants of *ragB* share only 43–56% of amino acids, which explains why the primers used are i) allele-specific and ii) designed on the most-conserved internal sequences, leading to only partial gene amplification [7]. In a previously published study, it was reported that primer efficiency across all gene polymorphisms of *ragAB* for clinical isolates may be impaired [43]. Considering that there are four different variants of *rag* [7], it is even more difficult to draw conclusions about the overall prevalence of *ragB* or different alleles from the currently available data.

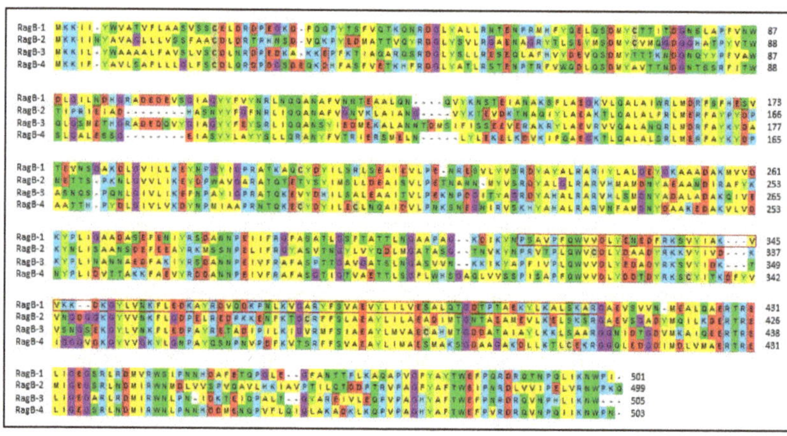

Figure 6. Amino-acid sequences of the different variants of RagB (types 1 to 4), as published by Hall et al., 2005 [7]. The relatively small part of RagB-1 investigated in the present study is outlined in red.

In view of the highly variable pathogenicity of *P. gingivalis* in general and *ragB* in particular, studies on the prevalence of the most virulent variant *ragB-1* are particularly important. The question remains of what function the conserved RagB-1 protein fragment has. In comparison with previously published 3D structural data for RagB [20,37], the sequence amplified in this study (positions 319 to 413) appears to represent part of one of the protein inserts (insert C) and α-helices nos. 12 to 15 of RagB-1.

Whether different RagB sequence variants contribute to functional variation remains to be investigated. A new hypothesis based on studies by Madej et al. [18] suggests that different RagABs enable *P.g.* strains to feed on different peptides produced during the degradation of host proteins [10,18]. The authors showed that the ligand-binding site of RagB-1 has an acidic character, which suggests that W83 preferentially takes up basic peptides. Interestingly, this acidic loop ($_{99}$DEDE$_{102}$) is absent in several RagB orthologs, including Rag-2 and Rag-4 [18], but is also found in a truncated form ($_{99}$DED$_{101}$) in RagB-3. Studies have shown that the growth of strains with different RagAB variants is identical on rich medium, whereas on minimal medium with bovine serum albumin (BSA) as the sole carbon source, robust growth was only observed for W83 (RagAB-1), whereas ATCC 33277 (RagAB-4) only grew slowly in contrast [7,18]. In the study by Madej et al. [18], an ATCC 33277 strain was constructed in which *ragAB-4* (or solely *ragB-4)* was replaced by *ragAB-1* (or solely *ragB-1)* from W83. Remarkably, replacement resulted in robust growth of ATCC 33277 on BSA. These results are significant: First, they confirm that the RagAB type does influence the growth of *P. gingivalis* on extracellular protein-derived oligopeptides. They suggest, secondly, that different RagB lipoproteins can form functional complexes with the same RagA transporter and, thirdly, that RagB appears to determine the substrate specificity of the complex.

In addition, the question remains open of where exactly the substrate receptor of the RagAB complex is located. The fact that studies have shown that one and the same RagA protein can bind and transport different substrates by recombination with different RagB "lids" [18] suggests that the substrate specificity of the RagAB complex is determined by RagB. It therefore seems plausible that the "receptor" for this transport is also encoded by the *ragB* gene. Goulas et al. [20] reported that, in most TPRPs, ligand binding occurs at the concave solenoid surface, whereas in *B. thetaiotaomicron,* the *inserts* form the binding site. These inserts differ significantly in sequence and trajectory, which contributes to *B. thetaiotaomicron* having a rather flexible binding site that facilitates binding to oligosaccharide molecules by recognizing the overall 3D shape, rather than the exact composition of the individual monosaccharides [20,47,48].

Because of the sequence homology to SusD of *B. thetaiotaomicron*, which is involved in saccharide uptake, and because RagB of *P. gingivalis* has putative saccharide-binding sites as well, it was originally assumed that its function included saccharide transport [10,14]. However, large differences in inserts A, B, and C indicate that the glycan-binding site of SusD is missing in RagB [20]. Goulas et al. [20] also draw parallels to NanU, where, due to significant differences in the regions shaping the sugar-binding site in SusD, the sialic acid-binding site is still unknown [20,49]. Since studies have now shown that RagAB is indeed a transport system for proteinaceous nutrients [18], the question remains as to why these saccharide-binding sites and bound monosaccharides are present on the RagB protein. Originally, this protein may have served as a saccharide transporter and, after transfer to *P.g.*, gradually converted into a protein transporter, with the saccharide-binding sites remaining as a "remnant" on the surface. Of course, the process could also have taken place in reverse, with the protein complex, originally used as a peptide transporter, being gradually converted into a saccharide transporter after acquisition by *B. thetaiotaomicron*. In conclusion here, however, the question of the origin or remaining function of the saccharide-binding sites and the bound monosaccharides on the surface of RagB remains unanswered.

The genes of the *rag* locus are part of the core genome of *P.g.* but have been reported to be highly variable between strains [43,50]. Nevertheless, the part of the *ragB-1* gene examined in this study was found to be remarkably conserved, and if mutations occurred, they were either synonymous (possibly silent) or at least neutral. On the basis of this high conservation of the section of the *ragB-1* gene examined, it can be assumed to be particularly important. The only two strains with non-synonymous mutations, originally classified as *P.g.*, were identified as *P. gulae* by Fournier et al. in 2001 [38] and are isolates from monkeys. Thus, *P. gulae* is an animal biotype of *P. gingivalis* and—unlike human *P.g.* strains—these two strains are catalase-positive.

Since the virulent *ragB-1* type is also found in animals, it can be speculated i) that *ragAB-1* was probably transferred from animals to humans, ii) that this probably occurred by a gum infection, and iii) that the virulent variant was weakened (tamed) by mutations over time so that new, less virulent, adapted commensal versions of *ragAB*, and thus *P.g.*, have evolved. Further studies are needed in order to elucidate the origin and significance of the *ragAB* locus and its variants, potentially useful as a diagnostic or prognostic marker for *P.g.*-associated periodontitis, and to establish a correlation between structural and functional differences.

5. Conclusions

A total of 22 of the 138 *P.g.* strains investigated in this study tested positive for a 436 bp conserved virulence-associated fragment of *ragB-1*, which corresponds to a prevalence of 16%. Variations in the base sequence were only detected in three strains (OMI 1072, OMI 1081, and OMI 1074). In addition, these variations were either synonymous or neutral. From this high conservation, it can be assumed that this section of the *ragB-1* gene must code for an essential function. Since the virulent *ragB-1* type is also found in animals, we speculate that *ragAB-1* was probably transferred from *P. gulae* animal strains to *P. gingivalis* human strains by horizontal gene transfer (as similarly demonstrated for the *fimA* locus encoding *P. gingivalis* long fimbrial stalk protein [51]) and that this originally virulent variant became less virulent by mutations over time as a prerequisite for commensalistic coexistence.

Author Contributions: Conceptualization, G.C.; methodology, G.C. and E.D.; software, H.L.M. and S.B.; acquisition, analysis, and interpretation of data, G.C., E.D., H.L.M. and S.B; validation, G.C. and S.B.; formal analysis, G.C. and S.B.; investigation, G.C. and S.B.; resources, G.C.; writing—original draft preparation, S.B.; writing—review and editing, G.C., H.L.M. and E.D.; visualization, S.B.; supervision, G.C.; project administration, G.C.; funding acquisition, G.C. All authors have read and agreed to the published version of the manuscript.

Funding: This research was funded by the Interdisciplinary Center for Clinical Research (IZKF) at the RWTH Aachen medical faculty, grant number OC1-6.

Institutional Review Board Statement: Not applicable.

Informed Consent Statement: Not applicable.

Data Availability Statement: The data presented in this study are available on request from the corresponding author S.B. and/or senior author G.C.

Acknowledgments: Our thanks go to Beate Melzer-Krick for her excellent technical assistance.

Conflicts of Interest: The authors hereby declare that there were no conflict of interest in the preparation of this article.

References

1. Henry, L.G.; McKenzie, R.M.E.; Robles, A.; Fletcher, H.M. Oxidative stress resistance in *Porphyromonas gingivalis*. *Future Microbiol.* **2012**, *7*, 497–512. [CrossRef]
2. Dahlén, G.; Gmür, R.; Yoshino, T. Phenotypes, serotypes and antibiotic susceptibility of Swedish *Porphyromonas gingivalis* isolates from periodontitis and periodontal abscesses. *Oral Microbiol. Immunol.* **2007**, *22*, 80–86. [CrossRef] [PubMed]
3. Amano, A.; Nakagawa, I.; Kataoka, K.; Morisaki, I.; Hamada, S. Distribution of *Porphyromonas gingivalis* strains with fimA genotypes in periodontitis patients. *J. Clin. Microbiol.* **1999**, *37*, 1426–1430. [CrossRef]
4. Griffen, A.L.; Lyons, S.R.; Becker, M.R.; Moeschberger, M.L.; Leys, E.J. *Porphyromonas gingivalis* strain variability and periodontitis. *J. Clin. Microbiol.* **1999**, *37*, 4028–4033. [CrossRef]
5. Missailidis, C.G.; Umeda, J.E.; Ota-Tsuzuki, C.; Anzai, D.; Mayer, M.P.A. Distribution of fimA genotypes of *Porphyromonas gingivalis* in subjects with various periodontal conditions. *Oral Microbiol. Immunol.* **2004**, *19*, 224–229. [CrossRef] [PubMed]
6. Nakagawa, I.; Amano, A.; Ohara-Nemoto, Y.; Endoh, N.; Morisaki, I.; Kimura, S.; Kawabata, S.; Hamada, S. Identification of a new variant of fimA gene of *Porphyromonas gingivalis* and its distribution in adults and disabled populations with periodontitis. *J. Periodontal Res.* **2002**, *37*, 425–432. [CrossRef] [PubMed]
7. Hall, L.M.C.; Fawell, S.C.; Shi, X.; Faray-Kele, M.-C.; Aduse-Opoku, J.; Whiley, R.A.; Curtis, M.A. Sequence diversity and antigenic variation at the rag locus of *Porphyromonas gingivalis*. *Infect. Immun.* **2005**, *73*, 4253–4262. [CrossRef]
8. Frandsen, E.V.; Poulsen, K.; Curtis, M.A.; Kilian, M. Evidence of recombination in *Porphyromonas gingivalis* and random distribution of putative virulence markers. *Infect. Immun.* **2001**, *69*, 4479–4485. [CrossRef]
9. Sawada, K.; Kokeguchi, S.; Hongyo, H.; Sawada, S.; Miyamoto, M.; Maeda, H.; Nishimura, F.; Takashiba, S.; Murayama, Y. Identification by subtractive hybridization of a novel insertion sequence specific for virulent strains of *Porphyromonas gingivalis*. *Infect. Immun.* **1999**, *67*, 5621–5625. [CrossRef] [PubMed]

10. Potempa, J.; Madej, M.; Scott, D.A. The RagA and RagB proteins of *Porphyromonas gingivalis*. *Mol. Oral Microbiol.* **2021**, *36*, 225–232. [CrossRef]
11. Kuhn, C. Genetische Polymorphismen in *Porphyromonas gingivalis* und deren Auswirkung auf dessen Virulenz. Ph.D. Dissertation, Heinrich-Heine-Universität, Düsseldorf, Germany, 2018.
12. Nagano, K.; Murakami, Y.; Nishikawa, K.; Sakakibara, J.; Shimozato, K.; Yoshimura, F. Characterization of RagA and RagB in *Porphyromonas gingivalis*: Study using gene-deletion mutants. *J. Med. Microbiol.* **2007**, *56*, 1536–1548. [CrossRef]
13. Bonass, W.A.; Marsh, P.D.; Percival, R.S.; Aduse-Opoku, J.; Hanley, S.A.; Devine, D.A.; Curtis, M.A. Identification of ragAB as a temperature-regulated operon of *Porphyromonas gingivalis* W50 using differential display of randomly primed RNA. *Infect. Immun.* **2000**, *68*, 4012–4017. [CrossRef]
14. Curtis, M.A.; Hanley, S.A.; Aduse-Opoku, J. The *rag* locus of *Porphyromonas gingivalis*: A novel pathogenicity island. *J. Periodontal Res.* **1999**, *34*, 400–405. [CrossRef] [PubMed]
15. Masuda, T.; Murakami, Y.; Noguchi, T.; Yoshimura, F. Effects of various growth conditions in a chemostat on expression of virulence factors in *Porphyromonas gingivalis*. *Appl. Environ. Microbiol.* **2006**, *72*, 3458–3467. [CrossRef] [PubMed]
16. Hanley, S.A.; Aduse-Opoku, J.; Curtis, M.A. A 55-kilodalton immunodominant antigen of *Porphyromonas gingivalis* W50 has arisen via horizontal gene transfer. *Infect. Immun.* **1999**, *67*, 1157–1171. [CrossRef] [PubMed]
17. Shi, X.; Hanley, S.A.; Faray-Kele, M.-C.; Fawell, S.C.; Aduse-Opoku, J.; Whiley, R.A.; Curtis, M.A.; Hall, L.M.C. The rag locus of *Porphyromonas gingivalis* contributes to virulence in a murine model of soft tissue destruction. *Infect. Immun.* **2007**, *75*, 2071–2074. [CrossRef]
18. Madej, M.; White, J.B.R.; Nowakowska, Z.; Rawson, S.; Scavenius, C.; Enghild, J.J.; Bereta, G.P.; Pothula, K.; Kleinekathoefer, U.; Baslé, A.; et al. Structural and functional insights into oligopeptide acquisition by the RagAB transporter from *Porphyromonas gingivalis*. *Nat. Microbiol.* **2020**, *5*, 1016–1025. [CrossRef] [PubMed]
19. Glenwright, A.J.; Pothula, K.R.; Bhamidimarri, S.P.; Chorev, D.S.; Baslé, A.; Firbank, S.J.; Zheng, H.; Robinson, C.V.; Winterhalter, M.; Kleinekathöfer, U.; et al. Structural basis for nutrient acquisition by dominant members of the human gut microbiota. *Nature* **2017**, *541*, 407–411. [CrossRef]
20. Goulas, T.; Garcia-Ferrer, I.; Hutcherson, J.A.; Potempa, B.A.; Potempa, J.; Scott, D.A.; Gomis-Rüth, F.X. Structure of RagB, a major immunodominant outer-membrane surface receptor antigen of *Porphyromonas gingivalis*. *Mol. Oral Microbiol.* **2016**, *31*, 472–485. [CrossRef]
21. Curtis, M.A.; Slaney, J.M.; Carman, R.J.; Johnson, N.W. Identification of the major surface protein antigens of *Porphyromonas gingivalis* using IgG antibody reactivity of periodontal case-control serum. *Oral Microbiol. Immunol.* **1991**, *6*, 321–326. [CrossRef]
22. Imai, M.; Murakami, Y.; Nagano, K.; Nakamura, H.; Yoshimura, F. Major outer membrane proteins from *Porphyromonas gingivalis*: Strain variation, distribution, and clinical significance in periradicular lesions. *Eur. J. Oral Sci.* **2005**, *113*, 391–399. [CrossRef] [PubMed]
23. Zeller, I.; Hutcherson, J.A.; Lamont, R.J.; Demuth, D.R.; Gumus, P.; Nizam, N.; Buduneli, N.; Scott, D.A. Altered antigenic profiling and infectivity of *Porphyromonas gingivalis* in smokers and non-smokers with periodontitis. *J. Periodontol.* **2014**, *85*, 837–844. [CrossRef] [PubMed]
24. Hutcherson, J.A.; Bagaitkar, J.; Nagano, K.; Yoshimura, F.; Wang, H.; Scott, D.A. *Porphyromonas gingivalis* RagB is a proinflammatory signal transducer and activator of transcription 4 agonist. *Mol. Oral Microbiol.* **2015**, *30*, 242–252. [CrossRef] [PubMed]
25. Liu, Y.; Zhang, Y.; Wang, L.; Guo, Y.; Xiao, S. Prevalence of *Porphyromonas gingivalis* four *rag* locus genotypes in patients of orthodontic gingivitis and periodontitis. *PLoS ONE* **2013**, *8*, e61028. [CrossRef]
26. Nelson, K.E.; Fleischmann, R.D.; DeBoy, R.T.; Paulsen, I.T.; Fouts, D.E.; Eisen, J.A.; Daugherty, S.C.; Dodson, R.J.; Durkin, A.S.; Gwinn, M.; et al. Complete genome sequence of the oral pathogenic bacterium *Porphyromonas gingivalis* strain W83. *J. Bacteriol.* **2003**, *185*, 5591–5601. [CrossRef]
27. Xu, J.; Bjursell, M.K.; Himrod, J.; Deng, S.; Carmichael, L.K.; Chiang, H.C.; Hooper, L.V.; Gordon, J.I. A genomic view of the human-*Bacteroides thetaiotaomicron* symbiosis. *Science* **2003**, *299*, 2074–2076. [CrossRef]
28. Chen, T.; Hosogi, Y.; Nishikawa, K.; Abbey, K.; Fleischmann, R.D.; Walling, J.; Duncan, M.J. Comparative whole-genome analysis of virulent and avirulent strains of *Porphyromonas gingivalis*. *J. Bacteriol.* **2004**, *186*, 5473–5479. [CrossRef] [PubMed]
29. Roberts, A.P.; Pratten, J.; Wilson, M.; Mullany, P. Transfer of a conjugative transposon, Tn5397 in a model oral biofilm. *FEMS Microbiol. Lett.* **1999**, *177*, 63–66. [CrossRef]
30. Waters, V.L. Conjugative transfer in the dissemination of beta-lactam and aminoglycoside resistance. *Front. Biosci.* **1999**, *4*, D433–D456. [CrossRef]
31. Li, Y.H.; Lau, P.C.; Lee, J.H.; Ellen, R.P.; Cvitkovitch, D.G. Natural genetic transformation of *Streptococcus mutans* growing in biofilms. *J. Bacteriol.* **2001**, *183*, 897–908. [CrossRef]
32. Li, Y.-H.; Tang, N.; Aspiras, M.B.; Lau, P.C.Y.; Lee, J.H.; Ellen, R.P.; Cvitkovitch, D.G. A quorum-sensing signaling system essential for genetic competence in *Streptococcus mutans* is involved in biofilm formation. *J. Bacteriol.* **2002**, *184*, 2699–2708. [CrossRef] [PubMed]
33. Wang, B.Y.; Chi, B.; Kuramitsu, H.K. Genetic exchange between *Treponema denticola* and *Streptococcus gordonii* in biofilms. *Oral Microbiol. Immunol.* **2002**, *17*, 108–112. [CrossRef] [PubMed]

34. Stork, M.; Bos, M.P.; Jongerius, I.; de Kok, N.; Schilders, I.; Weynants, V.E.; Poolman, J.T.; Tommassen, J. An outer membrane receptor of *Neisseria meningitidis* involved in zinc acquisition with vaccine potential. *PLoS Pathog.* **2010**, *6*, e1000969. [CrossRef] [PubMed]
35. Kong, F.; Zheng, D.; She, P.; Ni, P.; Zhu, H.; Xu, H.; Su, Z. *Porphyromonas gingivalis* B cell Antigen Epitope Vaccine, pIRES-ragB′-mGITRL, Promoted RagB-Specific Antibody Production and Tfh Cells Expansion. *Scand. J. Immunol.* **2015**, *81*, 476–482. [CrossRef] [PubMed]
36. Zheng, D.; Sun, Q.; Su, Z.; Kong, F.; Shi, X.; Tong, J.; Shen, P.; Peng, T.; Wang, S.; Xu, H. Enhancing specific-antibody production to the ragB vaccine with GITRL that expand Tfh, IFN-γ(+) T cells and attenuates *Porphyromonas gingivalis* infection in mice. *PLoS ONE* **2013**, *8*, e59604. [CrossRef]
37. National Center for Biotechnology Information. [Internet]. Bethesda (MD): National Library of Medicine (US), National Center for Biotechnology Information. 1998. Available online: https://www.ncbi.nlm.nih.gov/ (accessed on 19 March 2023).
38. Fournier, D.; Mouton, C.; Lapierre, P.; Kato, T.; Okuda, K.; Ménard, C. *Porphyromonas gulae* sp. nov., an anaerobic, gram-negative coccobacillus from the gingival sulcus of various animal hosts. *Int. J. Syst. Evol. Microbiol.* **2001**, *51*, 1179–1189. [CrossRef]
39. Hall, T.; Biosciences, I.; Carlsbad, C. BioEdit: An important software for molecular biology. *GERF Bull Biosci* **2011**, *2*, 60–61.
40. Hall, T.A. *BioEdit: A User-Friendly Biological Sequence Alignment Editor and Analysis Program for Windows 95/98/NT*; Nucleic Acids Symposium Series; Oxford University Press: Oxford, UK, 1999.
41. Tamura, K.; Stecher, G.; Kumar, S. MEGA11: Molecular Evolutionary Genetics Analysis Version 11. *Mol. Biol. Evol.* **2021**, *38*, 3022–3027. [CrossRef]
42. Dolgilevich, S.; Rafferty, B.; Luchinskaya, D.; Kozarov, E. Genomic comparison of invasive and rare non-invasive strains reveals *Porphyromonas gingivalis* genetic polymorphisms. *J. Oral Microbiol.* **2011**, *3*, 5764. [CrossRef]
43. Bunte, K.; Kuhn, C.; Walther, C.; Peters, U.; Aarabi, G.; Smeets, R.; Beikler, T. Clinical significance of *ragA*, *ragB*, and PG0982 genes in *Porphyromonas gingivalis* isolates from periodontitis patients. *Eur. J. Oral Sci.* **2021**, *129*, e12776. [CrossRef]
44. Su, Z.; Kong, F.; Wang, S.; Chen, J.; Yin, R.; Zhou, C.; Zhang, Y.; He, Z.; Shi, Y.; Xue, Y.; et al. The rag locus of *Porphyromonas gingivalis* might arise from *Bacteroides* via horizontal gene transfer. *Eur. J. Clin. Microbiol. Infect. Dis.* **2010**, *29*, 429–437. [CrossRef]
45. Dashper, S.G.; Mitchell, H.L.; Seers, C.A.; Gladman, S.L.; Seemann, T.; Bulach, D.M.; Chandry, P.S.; Cross, K.J.; Cleal, S.M.; Reynolds, E.C. *Porphyromonas gingivalis* Uses Specific Domain Rearrangements and Allelic Exchange to Generate Diversity in Surface Virulence Factors. *Front. Microbiol.* **2017**, *8*, 48. [CrossRef]
46. Laine, M.L.; van Winkelhoff, A.J. Virulence of six capsular serotypes of *Porphyromonas gingivalis* in a mouse model. *Oral Microbiol. Immunol.* **1998**, *13*, 322–325. [CrossRef] [PubMed]
47. Anderson, K.L.; Salyers, A.A. Genetic evidence that outer membrane binding of starch is required for starch utilization by *Bacteroides thetaiotaomicron*. *J. Bacteriol.* **1989**, *171*, 3199–3204. [CrossRef] [PubMed]
48. Koropatkin, N.M.; Martens, E.C.; Gordon, J.I.; Smith, T.J. Starch catabolism by a prominent human gut symbiont is directed by the recognition of amylose helices. *Structure* **2008**, *16*, 1105–1115. [CrossRef] [PubMed]
49. Phansopa, C.; Roy, S.; Rafferty, J.B.; Douglas, C.W.I.; Pandhal, J.; Wright, P.C.; Kelly, D.J.; Stafford, G.P. Structural and functional characterization of NanU, a novel high-affinity sialic acid-inducible binding protein of oral and gut-dwelling Bacteroidetes species. *Biochem. J.* **2014**, *458*, 499–511. [CrossRef] [PubMed]
50. Brunner, J.; Wittink, F.R.A.; Jonker, M.J.; de Jong, M.; Breit, T.M.; Laine, M.L.; Soet, J.J.d.; Crielaard, W. The core genome of the anaerobic oral pathogenic bacterium *Porphyromonas gingivalis*. *BMC Microbiol.* **2010**, *10*, 252. [CrossRef] [PubMed]
51. Fujiwara-Takahashi, K.; Watanabe, T.; Shimogishi, M.; Shibasaki, M.; Umeda, M.; Izumi, Y.; Nakagawa, I. Phylogenetic diversity in *fim* and *mfa* gene clusters between *Porphyromonas gingivalis* and *Porphyromonas gulae*, as a potential cause of host specificity. *J. Oral Microbiol.* **2020**, *12*, 1775333. [CrossRef]

antibiotics

MDPI

Review

Application of Different Wavelengths of LED Lights in Antimicrobial Photodynamic Therapy for the Treatment of Periodontal Disease

Yasuo Takeuchi [1,*], Akira Aoki [2,*], Koichi Hiratsuka [3], Chanthoeun Chui [4], Akiko Ichinose [2,5], Nay Aung [6], Yutaro Kitanaka [7], Sakura Hayashi [2], Keita Toyoshima [2], Takanori Iwata [2] and Shinich Arakawa [1]

1 Department of Lifetime Oral Health Care Science, Graduate School of Medical and Dental Sciences, Tokyo Medical and Dental University (TMDU), Tokyo 113-8549, Japan; shinperi@tmd.ac.jp
2 Department of Periodontology, Graduate School of Medical and Dental Sciences, Tokyo Medical and Dental University (TMDU), Tokyo 113-8549, Japan; akiko.ichinose@hamadalab.com (A.I.); hayashi.peri@tmd.ac.jp (S.H.); toyoperi@tmd.ac.jp (K.T.); iwata.peri@tmd.ac.jp (T.I.)
3 Department of Biochemistry and Molecular Biology, Nihon University School of Dentistry at Matsudo, Chiba 271-8587, Japan; hiratsuka.koichi@nihon-u.ac.jp
4 Flora Dental Care, Phnom Penh 120407, Cambodia; lobkob@yahoo.com
5 Waseda Research Institute for Science and Engineering, Waseda University, Tokyo 169-8555, Japan
6 Laser Light Dental Clinic Periodontal and Implant Center, Yangon 11241, Myanmar; nayaung.mdy@gmail.com
7 Department of Oral Diagnosis and General Dentistry, Tokyo Medical and Dental University (TMDU), Tokyo 113-8549, Japan; kitanaka.peri@tmd.ac.jp
* Correspondence: takeuchi.peri@tmd.ac.jp (Y.T.); aoperi@tmd.ac.jp (A.A.)

Abstract: Therapeutic light has been increasingly used in clinical dentistry for surgical ablation, disinfection, bio-stimulation, reduction in inflammation, and promotion of wound healing. Photodynamic therapy (PDT), a type of phototherapy, has been used to selectively destroy tumor cells. Antimicrobial PDT (a-PDT) is used to inactivate causative bacteria in infectious oral diseases, such as periodontitis. Several studies have reported that this minimally invasive technique has favorable therapeutic outcomes with a low probability of adverse effects. PDT is based on the photochemical reaction between light, a photosensitizer, and oxygen, which affects its efficacy. Low-power lasers have been predominantly used in phototherapy for periodontal treatments, while light-emitting diodes (LEDs) have received considerable attention as a novel light source in recent years. LEDs can emit broad wavelengths of light, from infrared to ultraviolet, and the lower directivity of LED light appears to be suitable for plaque control over large and complex surfaces. In addition, LED devices are small, lightweight, and less expensive than lasers. Although limited evidence exists on LED-based a-PDT for periodontitis, a-PDT using red or blue LED light could be effective in attenuating bacteria associated with periodontal diseases. LEDs have the potential to provide a new direction for light therapy in periodontics.

Keywords: phototherapy; antimicrobial photodynamic therapy; periodontal disease; light-emitting diode

1. Introduction

Optical technology has been used in medicine for examining, diagnosing, and treating patients. Particularly in current dental treatments, it has various applications, such as surgical ablation, disinfection, bio-stimulation, suppression of inflammatory responses, and promotion of wound healing (Figure 1) [1]. Compared with conventional mechanical therapies, phototherapy, a minimally invasive treatment method, has the advantage of reducing pain, discomfort, and tissue damage. It also accelerates wound healing, thereby improving the quality of life. Antimicrobial photodynamic therapy (a-PDT) is a modern form of phototherapy, which has garnered attention as a new therapeutic method for eradicating oral bacterial infections, and its application in periodontal treatment has been investigated [2,3]. a-PDT has been predominantly performed using lasers for the treatment of periodontal infections, and its adjunctive effect when combined with non-surgical periodontal debridement has also been evaluated. Light-emitting diodes (LEDs) have been used as a new light source for a-PDT, and the number of studies on a-PDT using LEDs has increased in recent years. This review presents an overview of the light application in periodontal treatments and describes the potential application of different wavelengths of LED as a new light source in a-PDT/phototherapy for the management of periodontal diseases.

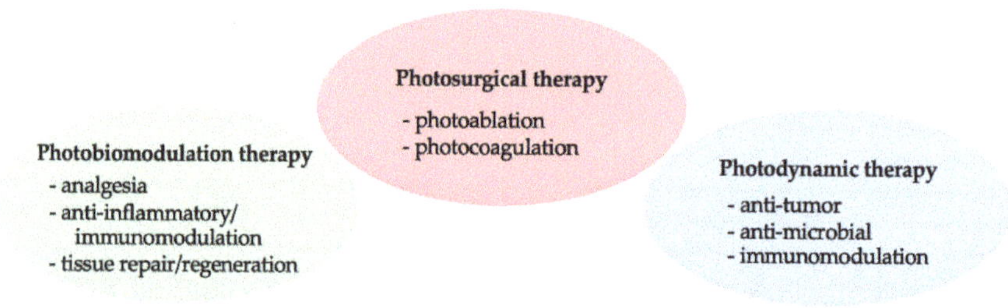

Figure 1. Various therapeutic effects obtained with the application of light.

2. Periodontal Diseases

Periodontal disease is an inflammatory condition affecting the periodontium, which leads to the destruction of periodontal tissues and eventual tooth loss. Approximately 1.1 billion individuals worldwide are reportedly affected by severe periodontal disease [4]. Recent epidemiological, clinical, and animal model studies have suggested a possible link between periodontal disease and various systemic diseases, such as diabetes and cardiovascular disease. Preventing and controlling periodontal disease is essential for maintaining oral and systemic health [5–7]. Although various environmental factors and systemic conditions may be involved in the onset and progression of diseases, bacterial plaque is generally considered the primary etiological factor. Mechanical instrumentation is considered in initial periodontal therapy for eliminating plaque and calculus deposits. However, conventional mechanical therapy may not entirely eliminate these etiologic factors in areas with limited access, such as deep pockets and complicated root surfaces. Antibiotic and antiseptic chemotherapies have limited efficacy against dental plaque residing as a biofilm [8,9], and the emergence of antibiotic-resistant bacteria is concerning [10,11].

3. Phototherapy in Periodontics

Optical technology is widely used for treating oral diseases, especially in the field of periodontics. Lasers, a monochromatic collimated beam of light, which can be concentrated into powerful energy beams within a narrow range, have been used as light sources for phototherapy in periodontics. High-power laser devices have been used to selectively cut hard and soft tissues and further remove diseased tissues. In particular, erbium-doped yttrium aluminum garnet (Er:YAG) laser is used to remove calculus in periodontal therapy [12,13]; the laser irradiation enables minimally invasive debridement and neutralization of diseased root surfaces. Photobiomodulation (PBM) uses low-power laser irradiation to trigger photophysical and photochemical reactions in host cells, resulting in beneficial physiological changes. Several studies have demonstrated the ability of PBM to promote rapid wound healing and alleviate pain at diseased sites [14–16]. Photodynamic therapy (PDT), which uses light and light-activatable photosensitizers, has garnered significant interest in recent years, with an increase in the number of basic and clinical studies on PDT.

PDT was originally developed as a treatment technique for cancers in various organs and age-related maculopathy [17,18]. PDT uses a specific wavelength of light to irradiate an area after administering a photosensitive substance, which has an affinity for the target tumor cells. The application of light excites the substance, and this process triggers the production of reactive oxygen species (ROS), leading to the destruction of the lesioned tissue. On the other hand, PDT used specifically to eradicate bacteria is known as "antimicrobial photodynamic therapy (a-PDT)". a-PDT has been used for the neutralization of harmful substances resulting from bacterial adherence to the root surfaces [19,20]. In vitro treatment with a-PDT potently and functionally inactivates inflammatory mediators, such as tumor necrosis factor (TNF)-α and interleukin (IL)-1β, which can impair periodontal restoration; thus, photochemical reactions from a-PDT may have the potential to enhance periodontal treatment [20]. The use of a-PDT has raised concerns regarding adverse effects on host cells. However, the light intensity used for bacterial elimination or suppression is below the toxicity limit of the host cell. Moreover, the light used during a-PDT exerts photobiological effects on the surrounding tissues, thereby providing important benefits for the periodontal healing process [1]. Although the mechanism is still debated, irradiated light reduces inflammation by regulating immunocompetent cells and supports the healing of periodontal tissues [21–29].

a-PDT is useful for sites, which are difficult to access with mechanical instruments. Several clinical studies have examined the effectiveness of a-PDT as an adjunctive therapy after conventional mechanical debridement of periodontal disease. Diode lasers have been used as light sources for a-PDT in the majority of studies. However, the discrepancies in the application conditions of a-PDT across studies have led to some researchers reporting inconsistent results, which limit the proven effectiveness of the adjunct use of a-PDT in the treatment of periodontal disease [30–32]. A 2021 systematic review by Moro et al. [33] evaluated 22 studies, which used a-PDT as an adjunctive therapy to scaling and root planing (SRP) with a 3-month follow-up. The results showed that the combination of a-PDT and SRP led to a significant increase in the clinical attachment level (CAL) and a decrease in the periodontal probing depth (PD) compared with those of SRP alone. They also found that the clinical efficacy of a-PDT was high when indocyanine green or high-concentration phenothiazine chloride was used as the photosensitizer. The beneficial effects of a-PDT also include reducing periodontopathic bacteria and improving clinical parameters [2].

Laser devices have been used as light sources in phototherapy for periodontal treatments. However, the activation of the photosensitizers used in a-PDT does not necessarily require high power light of lasers, and studies employing LEDs as light sources have increased in recent years. Similar to lasers, LEDs can be applied using a fiber-type tip to effectively irradiate anatomically complex areas, such as deep periodontal pockets. The lower directivity of LED light compared with that of lasers enables the irradiation of large and complicated surfaces using a flashlight-type device. Furthermore, LED devices are small, lightweight, and less expensive than lasers, making the production of home-use LED products for daily plaque control easier.

4. Antimicrobial Effects of Various Wavelengths of LEDs

LED devices can output wavelengths of light in the ultraviolet (UV), visible, and infrared regions. Several major periodontopathic bacteria possess endogenous photosensitizers, which are inactivated by the wavelength of LED light irradiation alone [34–36]. However, more than 700 bacterial species reside in the oral cavity, and the combined use of LED lights and photosensitizers is necessary to exert an antibacterial effect against various micro-organisms in clinical practice. The antibacterial effect of a-PDT is influenced by several factors, such as the dose and type of photosensitizer, wavelength and irradiance of light, and oxygen content in the irradiation field. A higher antibacterial effect can be expected when the wavelength of the irradiation light matches the absorption wavelength of the photosensitizer. Most photosensitizers can adhere to both bacteria and periodontal tissues and exhibit toxicity to the bacteria even in the absence of light [37]. The clinical application of a-PDT in periodontal treatment requires high antibacterial effects to be accompanied by minimal damage to host cells. The number of clinical studies investigating antibacterial phototherapy using LEDs is substantially lower than that of studies using lasers; however, in vitro studies examining the antibacterial effects of various wavelengths of LED lights with and without the use of photosensitizers have been reported increasingly.

4.1. Search Strategy

An electronic search of PubMed/MEDLINE, Cochrane Central Register of Controlled Trials (CENTRAL), and Web of Science databases was conducted for studies in March 2023. The search was limited to literature published in English. Literature searches in these databases were performed using the following terms: ("periodontal diseases" OR "periodontitis" OR "gingivitis" OR "periodontics" OR "periodontal") AND ("light-emitting diode" OR "light emitting diode" OR "LED") AND ("phototherapy" OR "photodynamic" OR "photodynamic therapy" OR "antimicrobial photodynamic therapy" OR "antibacterial photodynamic therapy" OR "a-PDT" OR "PDT"). An additional manual search of the references listed in all the included articles was conducted to identify further possible articles. The literature resulting from the search was screened by the authors first on the basis of titles and abstracts, and subsequently, on the basis of full texts. Particularly in clinical studies, the full-length articles of studies were evaluated in detail for inclusion in this review, and case reports/series, studies investigating the effect of LED light irradiation alone without photosensitizer, and studies targeting the treatment of peri-implant disease were excluded. Data extracted from the clinical research papers using LED-based a-PDT listed in Table 1 included publication information, study design, sample size, photosensitizer type/concentration, irradiation parameters and regimen, investigated parameters, and follow-up period.

4.2. Red Light/Infrared Light

The use of red light as an excitation light in a-PDT for periodontitis has been studied extensively. Phenothiazine-based photosensitizers, such as toluidine blue and methylene blue, show absorption peaks around 600–700 nm and are frequently used in combination with red light in a-PDT for periodontal diseases. The increased proportion of Gram-negative anaerobic bacteria in the dysbiotic microbiota of the diseased sites in the periodontium and the relatively small molecular weight and structural cationic charge of these photosensitizers facilitate easy penetration into the outer membrane of both Gram-negative and Gram-positive bacteria with a high affinity.

Red/infrared diode laser was used as the light source for a-PDT with blue dye initially, and its antibacterial effect against Gram-negative periodontopathic bacteria, such as *Porphyromonas gingivalis* and *Aggregatibacter actinomycetemcomitans*, has been demonstrated in vitro. Additional studies using red/infrared LED light have been conducted to confirm their efficacy [38,39]. Commercially available laser devices specifically designed for a-PDT in dental treatment were released before LEDs, and most clinical studies have used diode lasers. The clinical efficacy of a-PDT using red or near-infrared laser light with blue dye as an adjunctive therapy to mechanical debridement has been evaluated during the non-surgical and supportive phases of periodontal therapy. Wavelengths of 660 or 810 nm (radiant exposure: 0.6 to 1414.7 J/cm^2) were used with photosensitizers in these studies, and additional beneficial effects on clinical parameters and a reduction in the bacteriological burden were observed [2]. It has also been shown to be effective in combination with mechanical debridement during periodontal surgery [40].

Table 1. Clinical studies of LED-based antimicrobial photodynamic therapy in the treatment of periodontitis.

Reference	Study Participants (n) Study Design	Groups (n: Sites) Treatment Provided	Follow-Up Findings
Bassir et al., 2013 [41]	CP (16) Split-mouth	• SRP + a-PDT (119): SRP (US/H) +LED (625–635 nm, 2000 mW/cm^2, Fotosan®) +TB (0.1 mg/mL) Photosensitizer incubation time: not described Irradiation time: 10 s (inside) + 10 s (outside the pocket), ×6 (per tooth) • SRP + LED (96): SRP (US/H) +LED (625–635 nm, 2000 mW/cm^2, Fotosan®) Irradiation time: 10 s (inside) + 10 s (outside the pocket), ×6 (per tooth) • SRP + PS (90): SRP (US/H) + TB (0.1 mg/mL) • SRP (91): SRP (US/H)	• Follow-up: 3 months. • Adjunctive treatments (a-PDT, LED, or PS) were repeated 7 and 14 days later. • Photoactivation using LED did not show additional effects on clinical parameters (PD, CAL, BOP, PI) compared to SRP alone.
Pulikkotil et al., 2016 [42]	CP (16) Split-mouth	• SRP + a-PDT (16): SRP + LED (628 nm, Fotosan®) + MB Photosensitizer incubation time: 1 min Irradiation time: 10 s (inside) + 10 s (outside the pocket) • SRP (16): SRP	• Follow-up: 3 months. • Significantly greater improvement in BOP was seen in the SRP + a-PDT group compared to that in the SRP group after 3 months of treatment. • No difference in the quantity of Aa was detected between the groups.
Husejinagic et al., 2019 [43]	Periodontitis (20) Split-mouth, RCT	• SRP + a-PDT (20): SRP (US/H) +LED (635 nm, 750 mW, PADPLUS) + TB (12.7 μg/mL) Photosensitizer incubation time: 1 min Irradiation time: 10 s (inside the pocket), ×6 (per tooth) • SRP (20): SRP (US/H)	• Follow-up: 3 months. • Significant improvements in clinical parameters (PD, CAL, BOP) were shown in both groups, but the test and control groups were comparable. • The recolonization of Pg and Td was reduced after adjuvant treatment, but not significantly.

Table 1. *Cont.*

Reference	Study Participants (n) Study Design	Groups (n: Sites) Treatment Provided	Follow-Up Findings
Harmouche et al., 2019 [44]	Periodontitis (28) Split-mouth, RCT	• SRP + a-PDT (579): SRP (US/H) +LED (625–635 nm, 2000 mW/cm^2, Fotosan®) +TB (0.1 mg/mL) Photosensitizer incubation time: 1 min Irradiation time: 10–30 s (inside) + 10 s (outside the pocket) PDT applications were repeated 7 days and 3 months after SRP. • SRP (609): SRP (US/H)	• Follow-up: 6 months. • Repeated application of a-PDT with SRP significantly improved SRP outcome (PD and BOP) compared to SRP alone. • This effect was mainly observed at 6 months in initially deep sites (PD > 6 mm) with BOP.
Mongardini et al., 2014 [45]	CP (30); Residual pockets during SPT Split-mouth	• SRP + a-PDT (30): SRP (H) +LED (628 nm, 2000 mW/cm^2, Fotosan®) + TB (0.1 mg/mL) Photosensitizer incubation time: 1 min Irradiation time: 10 s (outside) + 10 s (inside the pocket), ×2 (per tooth) • SRP (30): SRP (H)	• Follow-up: 1 week. • One week after the treatment, the number of sites showing a PD reduction of ≥2 mm was higher in the a-PDT group than in the SRP group. • Higher reductions in relative proportions of red complex bacteria were observed in the a-PDT group compared to the SRP group.
Goh et al., 2017 [46]	Periodontitis (27); Residual pockets during SPT Split-mouth, RCT	• SRP + a-PDT (36): SRP (US/H) +LED (620–640 nm, 2000–4000 mW/cm^2, Fotosan®) +TB (0.1 mg/mL) Photosensitizer incubation time: not described Irradiation time: 20 s/site × 2 • SRP (36): SRP (US/H)	• Follow-up: 6 months. • At 3 months after treatment, significantly greater improvements in PD and CAL were observed in the SRP + a-PDT group compared to the SRP group. However, the differences were no longer significant at the 6-month follow-up. • Adjunctive a-PDT did not offer additional reduction in the levels of GCF cytokines, including IL-8, IL-6, and TNF-α.

Table 1. *Cont.*

Reference	Study Participants (n) Study Design	Groups (n: Sites) Treatment Provided	Follow-Up Findings
Hormdee et al., 2020 [47]	Periodontitis (12) Split-mouth, RCT	• SRP + a-PDT (12): SRP (US/H) +LED (420–480 nm, 1000–1200 mW/cm^2) +CUR gel (25 μg/mg) Photosensitizer incubation time: none (irradiated immediately) Irradiation time: 2 min (inside the pocket) per tooth • SRP (12): SRP (US/H)	• Follow-up: 6 weeks. • In the a-PDT group, significant reductions in PD and CAL were observed in the intragroup comparison from the first week up to the fourth week of follow-up. In contrast, a significant reduction in PD was observed after only a week in the SRP group. • The quantities of Fn and Pi were recovered in the SRP group, while there was no significant recolonization of these bacteria on PDT sites throughout the 6-week study duration.
Ivanaga et al., 2019 [48]	CP with type 2 DM (23); Residual pockets during SPT Split-mouth, RCT	• SRP + a-PDT (88): SRP (US/H) +LED (465–485 nm, 100 mW/cm^2, InGaN) + CUR (0.1 mg/mL) Irrigation with CUR for 1 min Irradiation time: 60 s (outside the pocket) • SRP + LED (80) SRP (US/H) +LED (465–485 nm, 100 mW/cm^2, InGaN), Irrigation with 1 mL saline solution Irradiation time: 60 s (outside the pocket) • SRP + PS (67): SRP (US/H) + CUR (0.1 mg/mL) • SRP (97): SRP (US/H) Treatment modalities were performed in a total of 332 sites, but only 290 sites were included in the final evaluation.	• Follow-up: 6 months. • Significant improvements in PD and BOP were shown in all treatment groups; however, the mean values for PD, CAL, GR, BOP, and PI did not differ among the four groups at baseline, 3-, and 6-month follow-ups. • Significant CAL gain was found only in the a-PDT and LED groups at 3 months in comparison to baseline data.

CP: chronic periodontitis, SRP: scaling root planing, H: hand scaling, US: ultrasonic scaling, a-PDT: antimicrobial photodynamic therapy, PS: photosensitizer, LED: light-emitting diode, MB: methylene blue, TB: toluidine blue, CUR: curcumin, PD: probing depth, CAL: clinical attachment level, GR: gingival recession, BOP: bleeding on probing, PI: plaque index, GI: gingival index, Aa: *Aggregatibacter actinomycetemcomitans*, Fn: *Fusobacterium nucleatum*, Pg: *Porphyromonas gingivalis*, Pi: *Prevotella intermedia*, Td: *Treponema denticola*, GCF: gingival crevicular fluid.

Conversely, a limited number of clinical studies have been conducted on a-PDT utilizing red LED light (Table 1). Most of these studies employed commercially available red LED devices (peak wavelength of 620–640 nm, 2000 mW/cm^2) with photosensitizers and similar irradiation conditions [20 J/cm^2/site × 2 from inside and outside the pocket, applying toluidine blue (0.1 mg/mL) prior to irradiation] and have demonstrated the efficacy of a-PDT as an adjunct to conventional mechanical treatment. However, several studies have reported conflicting results. For instance, Bassir et al. [41] examined the efficacy of LED-based a-PDT combined with conventional debridement in 16 patients with moderate-to-severe periodontitis and reported no additional benefits of a-PDT for the clinical outcomes compared with SRP alone. The discrepancies in the results may be due to differences in patient characteristics and a-PDT procedures (e.g., application time of photosensitizers and irradiation technique).

The morphology of the roots to which the bacteria are attached is complex. Periodontal pockets become deeper as periodontal disease progresses, making it increasingly difficult to align the light port with the target area even when a fiber-type tip is used. Red light is on the long-wavelength side of visible light and penetrates the gingiva to a greater extent. The penetration depth of light at the wavelength of 660 nm, which is commonly used in a-PDT, is approximately 3.0–3.5 mm [49]. Although light energy is attenuated as it passes through the gingiva [50,51], red light can act on the bacteria in the pockets, even when irradiated from the gingival surface. Transgingival irradiation facilitates easier and faster treatment without the insertion of the light into the periodontal pocket. In a study using a diode laser, Sasaki et al. [52] used a gingival model created using beef slices to determine the feasibility of transgingival a-PDT. They reported that a combination of indocyanine green-encapsulated nanoparticles (final concentration: 10 mg/mL) and infrared light from a diode laser (810 nm, 960–4800 J/cm^2) had an antimicrobial effect even with transgingival irradiation. Schär et al. [53] evaluated the clinical effects of transgingival irradiation with red diode laser light [670 nm, 39 or 58 J/cm^2 (single root tooth or molar)] and 1% methylene blue after SRP in periodontal pockets of patients with stage II–III periodontitis. They found that compared with the control group, the test group showed a significant reduction in bleeding on probing (BOP) and a trend toward improvement in PD and CAL after treatment. Compared with lasers, LED light is more suitable for irradiating a relatively wide area of the gingiva. The authors [54] also showed that a-PDT with red LED light (660 nm, 1.1 W/cm^2, 20 s/site) and toluidine blue (1 mg/mL) inhibited plaque formation. Wide-area irradiation using LEDs is easy to implement and can be applied for plaque control (Figure 2).

4.3. Blue Light

Short-wavelength lights, such as blue light, lack the ability to penetrate deeply into tissues; thus, they cannot be used as light sources for PDT, which targets tissues deep in the human body. However, fiber-shaped tips have enabled the direct irradiation of periodontal pockets, making it possible to apply short-wavelength light for a-PDT in periodontal treatment and supportive care. Blue light can be generated by diodes or argon lasers and has been used in clinical dentistry for bleaching and curing resins. However, the equipment cost limits its widespread use. In contrast, blue LED devices are less expensive than laser devices and are now widely used to cure resin materials used for restoring dental caries. Research on the benefits of blue LED light with the aim of applying it to phototherapy is ongoing.

Curcumin (maximum absorption wavelength: 425 nm), riboflavin (266 nm, 373 nm, 445 nm), rose bengal (550 nm), erythrosine (530 nm), and sinoporphyrin sodium (366 nm) have been used as photosensitizers in a-PDT using third-generation blue LED light [55]. The antibacterial effect of a-PDT in combination with blue LEDs and the aforementioned photosensitizers has been demonstrated in vitro on periodontal bacteria commonly found in patients with periodontitis, such as *A. actinomycetemcomitans*, *Campylobacter rectus*, *Fusobacterium nucleatum*, *Parvimonas micra*, *P. gingivalis*, *Tannerella forsythia*, *Prevotella intermedia*, and *Streptococcus gordonii* [56–65]. In addition, the antibacterial activity of blue light irradiation alone has been demonstrated against some oral bacteria (Figure 3) [36,66–71], further MRSA, and novel coronavirus [72,73]. Black-pigmented bacteria, including *P. gingivalis* and *P. intermedia*, which are commonly associated with periodontal diseases, degrade hemoglobin to derive heme, and endogenous porphyrin is produced when the bacteria further acquire iron from heme [74]. The importance of porphyrins in the antimicrobial action of blue light has been suggested [74]; irradiated blue light reacts with endogenous porphyrin as a photosensitizer, leading to the production of ROSs, which efficiently and selectively kill bacteria [3,34,57,68,75]. This characteristic could have clinical advantages, as blue light exposure without a photosensitizer suppresses only the growth of periodontal pathogenic bacteria possessing endogenous porphyrin while simultaneously minimizing the impact on the natural microbiome. Masson-Meyers et al. [76] compared the antibacterial effects of 405 nm LED and 405 nm diode laser irradiation with the same amount of energy against *Staphylococcus aureus* and reported that both light therapies significantly suppressed bacterial growth and that the differences between the two light sources did not affect the results. Although the phases of LED light are not aligned with those of laser light, the total energy may have a greater influence on the outcome in terms of antibacterial efficacy. Blue light is more readily absorbed by the human mucosal tissue than red light [49], and excessive exposure to high-power blue light can cause tissue damage. Therefore, it is necessary to carefully consider the irradiation conditions for clinical use, although direct irradiation of the pocket can minimize the effect on host cells, which is advantageous from a safety standpoint.

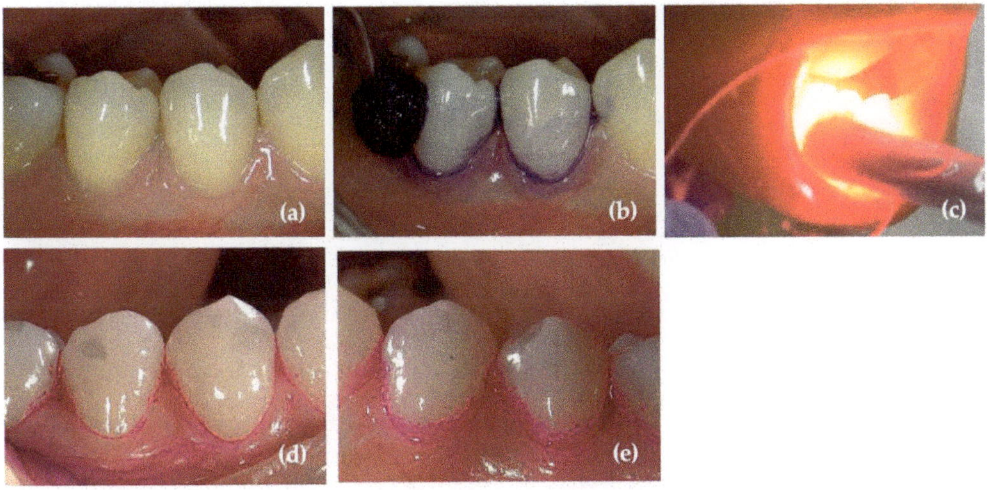

Figure 2. Suppression of dental plaque formation after antimicrobial photodynamic therapy (a-PDT) (red light-emitting diode (LED)/toluidine blue). Dental plaque was removed from tooth surfaces (**a**); toluidine blue O (1 mg/mL) was gently applied (**b**). After washing, red LED (660 nm, 1.1 W/cm^2) was focused on the tooth surfaces (**c**). Suppression of plaque formation on the a-PDT group teeth (**d**) could be confirmed compared to the control teeth (**e**). Figure from Ichinose-Tsuno, A et al. Antimicrobial photodynamic therapy suppresses dental plaque formation in healthy adults: a randomized controlled clinical trial. *BMC Oral Health* **2014**, 14, 152. CC-BY 4.0 [14].

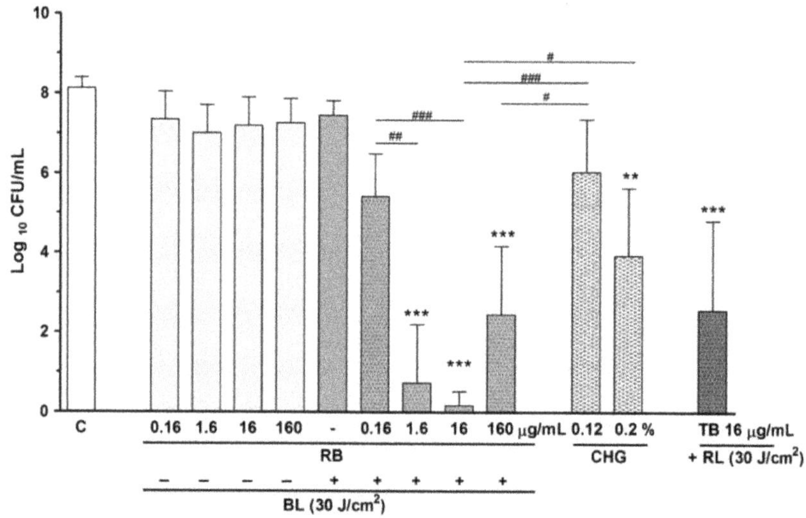

Figure 3. Effect of blue light-emitting diode (LED) (BL) in combination with graded rose bengal (RB). Fresh *Porphyromonas gingivalis* suspension was mixed with RB (the final concentrations were 0.16, 1.6, 16, and 160 μg/mL, respectively) and irradiated with 30 J/cm^2 BL. Bacterial suspensions treated with chlorhexidine gluconate (CHG) or other a-PDT [a combination of toluidine blue (TB) and red LED (RL) irradiation] were employed as the positive controls. After the treatment, each suspension was plated and incubated on brucella agar plates, and the numbers of colony-forming units (CFU) were determined. a-PDT using BL and RB showed a high antibacterial effect against periodontopathic bacteria in vitro. ** $p < 0.001$, *** $p < 0.0001$ (vs. Control). $^{\#}$ $p < 0.05$, $^{\#\#}$ $p < 0.001$, $^{\#\#\#}$ $p < 0.0001$. Figure from Chui, C et al. Antimicrobial effect of photodynamic therapy using high-power blue light-emitting diode and red-dye agent on *Porphyromonas gingivalis*. *J Periodontal Res* **2013**, 48, (6), 696–705. © Copyright (2013) John Wiley & Sons A/S. DOI: 10.1111/jre.12055 [36].

Only a few clinical studies have evaluated the effects of a-PDT using blue LED (Table 1). Hormdee et al. [47] investigated the efficacy of a-PDT using blue LED and *Curcuma longa* extract as adjuncts to SRP in the treatment of patients with moderate chronic periodontitis. The application of *Curcuma longa* gel (concentration: 25 μg/mg) after SRP followed by irradiation of the pockets with blue LED (16.8 J/cm^2) led to a significant improvement in PD and BOP at 4 weeks after treatment. The number of periodontopathic bacteria (*F. nucleatum* and *P. intermedia*) was also significantly suppressed after treatment. The probing depth decreased significantly one week after treatment in the control group; however, recurrence of the periodontal pocket with recolonization of periodontopathic bacteria was observed thereafter. Ivanaga et al. [48] investigated the clinical efficacy of a-PDT with curcumin solution (concentration: 0.1 g/L) and blue LED light (465–485 nm, 7.69 J/cm^2, irradiated outside of the pockets) as adjunctive therapy to SRP in residual pockets after periodontal therapy in patients with periodontal disease and type 2 diabetes. A reduction in PD and BOP positivity rates was observed in the SRP-only and a-PDT combined groups after treatment. However, significant CAL gain was observed only in the a-PDT combined group at three months in comparison to baseline data. Araújo et al. [77] used a mouthwash containing a curcumin solution (concentration: 1.5 g/L) and irradiated the mouths of 13 volunteers with blue LED (450 nm, 20.1 J/cm^2, irradiated the oral cavity by inserted light source) to determine the effectiveness of a-PDT for routine oral hygiene purposes. They examined the quantity of bacteria in the saliva before and after treatment and reported that a-PDT using curcumin mouthwash and blue LED significantly reduced the bacteria (68.3%), whereas the use of mouthwash alone did not result in an effective reduction in bacteria (9%). Ricci Donato et al. [78] instructed 50 volunteers to rinse with one of two photosensitizers (curcumin or a hematoporphyrin derivative). Light illumination (blue light at 450 nm for curcumin and red light at 630 nm for the hematoporphyrin derivative; irradiance: <100 W/cm^2 for 6 min) was performed using a LED device. The results showed a significant reduction in the number of bacteria in the saliva after both photodynamic treatments; a reduction was observed even after 24 h of treatment in the curcumin + blue light group. Genina et al. [79] investigated the efficacy of a toothbrush equipped with a blue light-emitting function (405–420 nm, 2 mW/cm^2) as a daily means of suppressing the progression of periodontal disease. Sixty participants with mild-to-moderate gingivitis were randomly divided into two groups (blue LED-emitting toothbrushes and regular toothbrushes). A reduction in the accumulation of dental plaque and gingival inflammation was observed in both groups after using the toothbrushes for one month; however, the efficiency (improvement in clinical indices) of brushing was significantly higher in the LED toothbrush group than in the control group.

Only a limited number of clinical studies have been conducted on a-PDT using blue LEDs, and there is heterogeneity in study designs. Thus, it is difficult to determine its clinical efficacy at present; however, many ongoing studies have suggested a growing potential for the clinical application of blue light.

4.4. Green/Yellow Light

Rose bengal (4,5,6,7-tetrachloro-2′,4′,5′,77-tetraiodofluorescein, RB) is present as a pigment in plaque-disclosing agents used in dentistry. a-PDT combined with RB as the photosensitizer and blue LED light (425–470 nm) exhibits high antibacterial activity against *P. gingivalis* in vitro (Figure 3) [58]. The maximum absorption wavelength of RB is approximately 550 nm. Therefore, Kitanaka et al. [80] investigated the antibacterial effects of a-PDT using a new combination of yellow-green LED light (565 nm) and RB against *P. gingivalis* in vitro. The results showed that a-PDT with yellow-green LED (8.56 J/cm^2) and RB showed higher antibacterial activity against *P. gingivalis* than a-PDT with blue LED light (470 nm, 8.56 J/cm^2) and RB. In addition, morphological changes suggesting leakage of bacterial contents were observed under a scanning electron microscope within a short period of time after a-PDT; no subsequent bacterial growth was observed (Figure 4). These results suggest that a-PDT with RB and yellow-green LED resulted in the physical destruction of the bacterial cell wall and high bactericidal activity. Green light has also been reported to have photobiomodulatory effects on osteoblasts and bone cells and has attracted considerable attention [81]. However, to the best of our knowledge, the effects of a-PDT using green or yellow-green wavelengths of light have not been clinically evaluated in dentistry, and this area of research is expected to progress in the future.

4.5. UV Light

The peak of light absorption for DNA is at 260 nm. UVC (200–280 nm) and UVB (280–320 nm) can induce the formation of pyrimidine-pyrimidone (6-4) photoproducts or cyclobutane-type pyrimidine dimers, which damage DNA [82–84]. The bactericidal effect of UV light on oral bacteria is thought to be mediated via a similar mechanism. In vitro studies have shown that UV-LED light has a bactericidal effect against oral bacteria, such as *F. nucleatum*, *P. gingivalis*, *S. mutans*, and *S. sanguinis* [85,86]. The antibacterial action of UV light alone (without the use of a photosensitizer) was evaluated in these studies (Figure 5). UVC-LED showed a strong bactericidal effect; however, high cytotoxicity was demonstrated in a cell viability test using gingival fibroblasts [86]. UVB-LED exhibited a weaker bactericidal effect on oral bacteria than UVC-LED; however, it showed lower cytotoxicity to gingival epithelial cells. UVB-LED may induce the production of ROS from oral epithelial cells and enhance bactericidal activity against specific periodontopathic bacteria, such as *P. gingivalis* [85]. These results suggest that UVB-LED can also be used to control infections in the oral cavity.

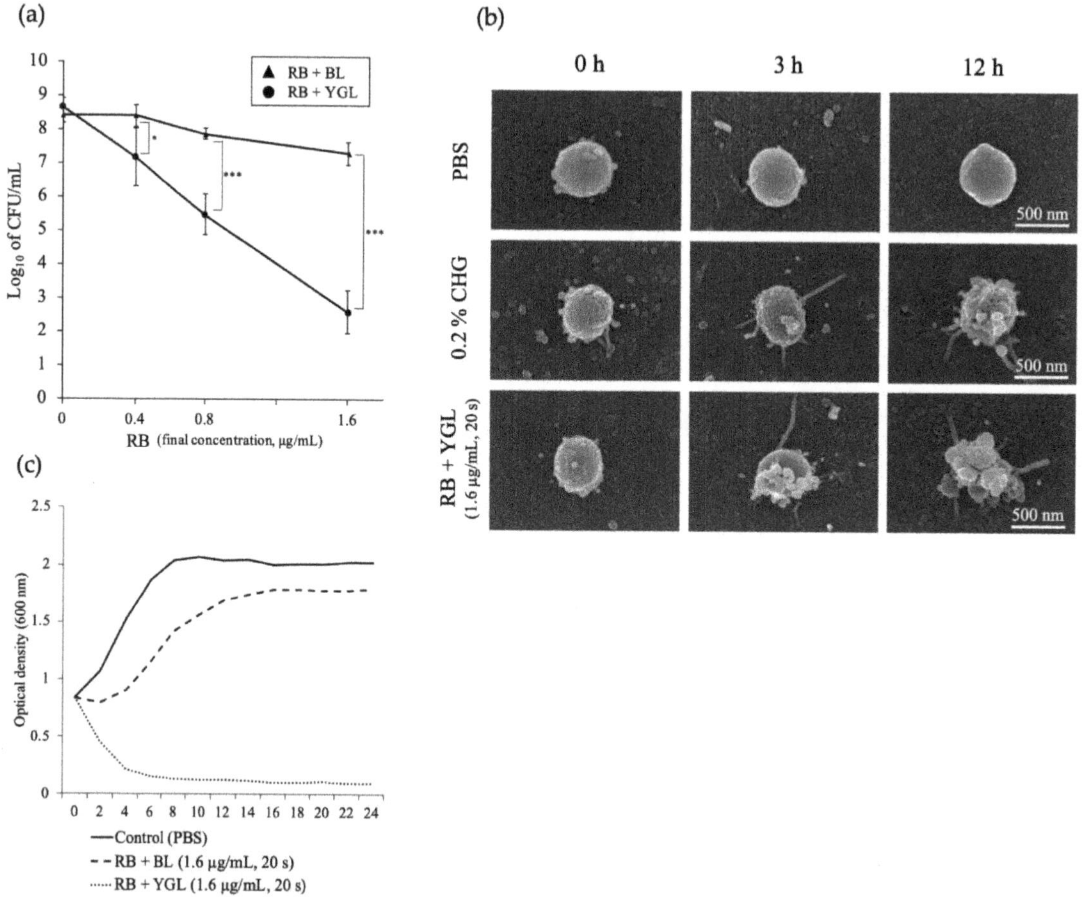

Figure 4. Antibacterial effects of a-PDT (rose bengal (RB) plus yellow-green light-emitting diode (LED) (YGL)) on *Porphyromonas gingivalis* (Pg). Pg suspension was mixed with RB and irradiated with YGL or BL for 20 s (8.56 J/cm^2) in vitro. (**a**) Antimicrobial photodynamic therapy (a-PDT) treatment employing RB + YGL significantly decreased viable bacteria as compared to RB + BL. (**b**) Morphological change in Pg was noticeable after a-PDT treatment (RB and YGL). (**c**) After the treatment with RB + BL, the Pg growth was inhibited temporarily and then increased to a plateau. In contrast, a dramatic reduction in Pg growth rate in the RB + YGL group was observed up to 6 h, finally reaching the lowest plateau. * $p < 0.05$, *** $p < 0.001$. Figure from Kitanaka, Y et al. The effect of antimicrobial photodynamic therapy using yellow-green LED and rose bengal on Porphyromonas gingivalis. *Photodiagnosis Photodyn Ther* **2020**, 32, 102033. © Copyright (2020) Elsevier. DOI: 10.1016/j.pdpdt.2020.102033 [80].

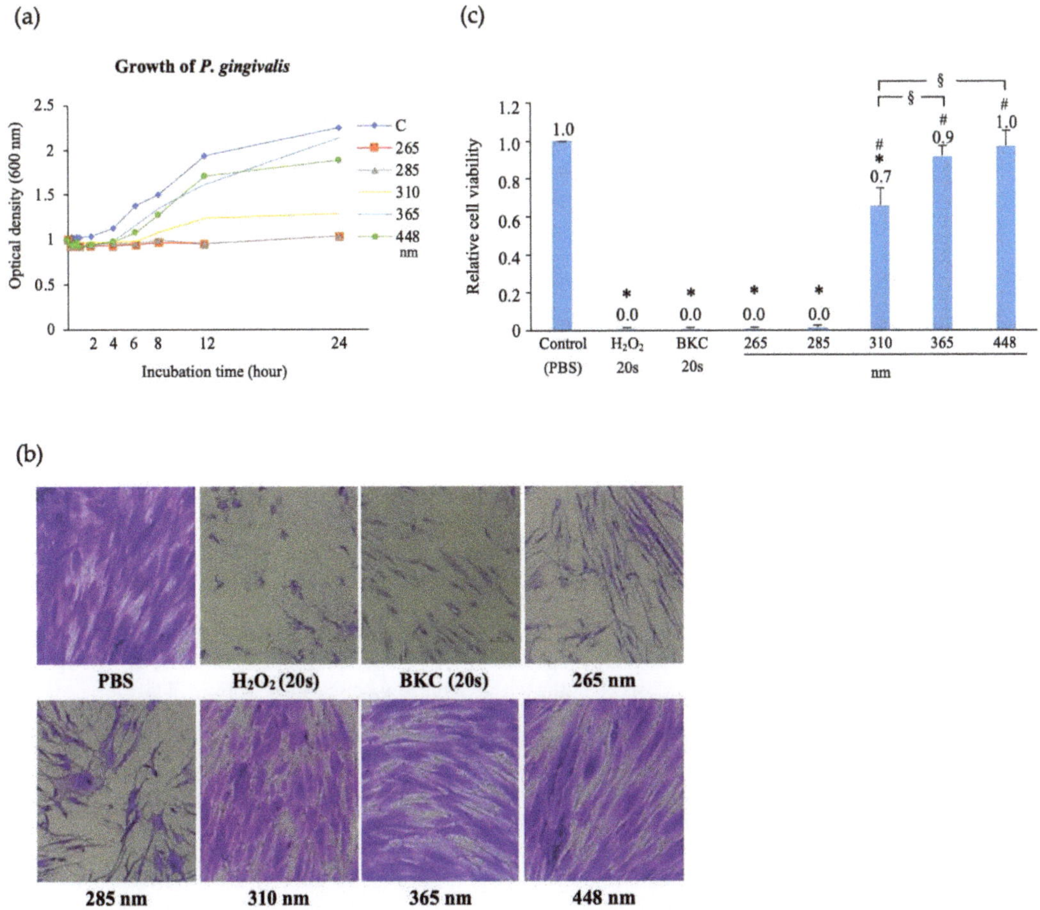

Figure 5. Effect of ultraviolet (UV) and blue light-emitting diode (LED) light on *Porphyromonas gingivalis* (Pg) and human gingival fibroblasts. (**a**) Growth of Pg after anaerobic irradiation (600 mJ/cm^2) with UV and blue LED light. Pg irradiated at 265 or 285 nm showed little growth over 24 h, while a slight growth increase was observed with 310 nm irradiation. (**b**) Crystal violet staining images of human gingival fibroblasts 24 h after UV-LED irradiation. (**c**) Viability of human gingival fibroblasts 24 h after UV-LED irradiations. Oral antiseptic agents [3% H_2O_2 (20 s) and 0.025% benzalkonium chloride (BKC) (20 s)] and phosphate-buffered saline (PBS) were used as controls. Irradiation with UVB light (265 or 285 nm, 60 s) completely devitalized HGF-1. Irradiation at 310 nm reduced viability by 30% after 24 h, whereas irradiations at 365 and 448 nm resulted in no significant reduction compared with control group. * $p < 0.05$ (vs. Control), $^\#$ $p < 0.05$ (vs. H_2O_2, BKC, 265, and 285 nm groups), § $p < 0.05$. Figure from Aung, N et al. The effects of ultraviolet light-emitting diodes with different wavelengths on periodontopathic bacteria in vitro. *Photobiomodul Photomed Laser Surg* **2019**, 37, (5), 288–297. © Copyright (2019) Mary Ann Liebert, Inc. DOI: 10.1089/photob.2018.4514 [86].

Narrow-band (NB)-UVB light, the specific wavelength around 310 nm with a narrow peak, has already been used to treat skin diseases—including psoriasis, vitiligo vulgaris, and atopic eczema—in dermatology. NB-UVB has fewer side effects on the host than broad-band-UVB light [87–89] and has been reported to upregulate regulatory T cells [90]. The immunosuppressive effects of NB-UVB may also be preferable for the treatment of periodontitis [91–94]. However, at present, few studies have investigated the application of

UVB-LED light in a-PDT or phototherapy for periodontal treatments, and the efficacy and safety should be carefully investigated through in vitro and clinical trials before applying it in clinical practice. UVB-induced immunosuppression may facilitate oral microbial infection. Over-irradiation with UVB light can cause damage to oral tissues and increase the risk of oral cancer [95,96].

5. Future Perspectives and Conclusions

Mechanical debridement at the site of infection remains the gold standard of treatment for periodontal diseases. However, phototherapy may be a useful adjunct to conventional therapy in anatomically complex areas and areas with limited access. a-PDT is a minimally invasive and virtually painless treatment, which does not produce antibiotic-resistant bacteria.

Although the irradiance of LEDs is generally low compared with that of lasers, a certain low threshold of light energy may be sufficient to induce bactericidal effects using a-PDT. LEDs generally have a low heat output; however, when used at high irradiance during treatment, the excessive heat generated by LEDs causes discomfort to patients. Moreover, there is a possibility of phototoxicity to neighboring tissues, which are part of the irradiation field. Therefore, further in vitro, in vivo, and clinical studies are required to determine the optimal irradiance, irradiation time, and other aspects to ensure the safety and effectiveness of LED phototherapy. To promote the clinical application of LED-based a-PDT, it is necessary to perform in vitro experiments under standardized irradiation conditions and establish recommended a-PDT protocols, which are presumed to be highly effective based on basic research data.

The wavelength of the irradiated light should be matched with the wavelength, which can be absorbed by the photosensitizer, particularly when performing a-PDT. LEDs produce light with various peak wavelengths within a narrow bandwidth. Thus, LEDs can provide light at an optimal wavelength for any photosensitizer (including endogenous photosensitizers in the pathogenic bacterial cells) used in phototherapy. However, simply increasing the dose of light or the concentration of a photosensitizer may not improve the effect of a-PDT [97,98]. Photochemical reactions and the characteristics of photosensitizers (i.e., the toxicity of the photosensitizer itself) also affect the results of a-PDT. A higher concentration of the dye can cause dimerization or multimerization of the photosensitizer and shift the absorption wavelength peak of the dye. The mismatch between the wavelengths of the light and the photosensitizer limits the production of ROS, resulting in attenuation of the antibacterial effect of a-PDT. At present, no consensus exists on the optimal setting of photosensitizers in a-PDT in periodontal practice (i.e., concentration, incubation time, etc.). Indeed, various concentrations of photosensitizers have been used for periodontal treatment of a-PDT (i.e., methylene blue: 0.005–1%; toluidine blue: 0.1–0.5%; and indocyanine green: 0.025–0.5%) [2]. Meanwhile, in most of the previous studies, the incubation time of the photosensitizers was set to 1 min, and the antibacterial efficacy of a-PDT was evaluated. A longer incubation time would increase the amounts of photosensitizers on the bacterial cells, resulting in strong antibacterial effects. However, the cytotoxicity of photosensitizers themselves has also been demonstrated in in vitro studies, and the incubation period would preferably be the minimum time required for the photosensitizers to be deposited on the target bacteria. Further in vitro and clinical studies should be conducted to determine this.

Inflammatory periodontal destruction is caused by dysbiotic polymicrobial communities, and at least 30–100 species are generally identified from a single periodontal pocket [99]. Recent research on the periodontal microbiota has clarified how the interactions between specialized community members determine an emergent overall function, which promotes or destabilizes periodontal tissue homeostasis. For example, although they are present in low numbers, keystone pathogens contribute to the emergence of dysbiotic microbiota by subverting the host immune response [100]. The appropriate wavelength of light, characteristics of the photosensitizer, and dose of light required to kill the bacteria vary with the species, and it may be necessary to determine the species or specific group of bacteria, which will be the therapeutic target of a-PDT.

Increasing the bactericidal effect is not the only way to enhance the therapeutic efficacy of a-PDT. Some wavelengths of light can activate host cells; this is known as PBM therapy [101–103]. The combination of multiple narrow wavelengths and intensities of LED light may be used as a new light source with high antibacterial properties and low toxicity to host cells, thereby increasing the effectiveness of periodontal treatment [104]. In addition, concurrent mechanical debridement of plaque biofilms and dental calculus will increase the effectiveness of a-PDT in treating periodontal disease. Bacteria surrounded by a biofilm matrix are less active and can withstand starvation and other harsh conditions. It is difficult for photosensitizers to penetrate biofilms, and in vitro studies have reported that biofilm-formed bacteria are more resistant to a-PDT than planktonic bacteria [35,105–107]. Although clinical validation is required, repeated application of a-PDT may also be effective [108,109].

The number of clinical trials using LED-based a-PDT for periodontitis is limited. Moreover, there is also heterogeneity in the study designs, which makes it difficult to compare their results. Nevertheless, previous research suggests that a-PDT using red or blue LED light appears effective in attenuating bacteria associated with periodontal diseases. Further randomized controlled trials with a standardized application protocol should be conducted to promote the clinical use of LED-based a-PDT. Progress in this research will increase the precision and reliability of these therapies in periodontal practice. LED devices have attractive features, which lasers do not possess. Thus, LEDs have the potential to provide a new direction for light therapy in periodontics.

Author Contributions: Conceptualization, Y.T. and A.A.; investigation, C.C., A.I., N.A., Y.K., S.H. and K.T.; data curation, Y.T. and A.A.; writing—original draft preparation, Y.T. and A.A.; writing—review and editing, Y.T., A.A., K.H., C.C., A.I., N.A., Y.K., S.H., K.T., T.I. and A.A.; visualization, Y.T.; supervision, A.A., T.I. and S.A.; project administration, Y.T. and A.A. All authors have read and agreed to the published version of the manuscript.

Funding: This work was supported by the Japan Society for the Promotion of Science KAKENHI (grant number 20K09971 to A.A. and Y.T., 20K09934 to Y.T.).

Institutional Review Board Statement: Not applicable.

Informed Consent Statement: Not applicable.

Data Availability Statement: No new data were created or analyzed in this study. Data sharing is not applicable to this article.

Acknowledgments: We greatly appreciate Yuichi Izumi (Southern TOHOKU Research Institute for Neuroscience, Southern TOHOKU General Hospital) for their excellent assistance.

Conflicts of Interest: The authors have no conflict of interest directly relevant to the content of this article.

References

1. Aoki, A.; Mizutani, K.; Schwarz, F.; Sculean, A.; Yukna, R.A.; Takasaki, A.A.; Romanos, G.E.; Taniguchi, Y.; Sasaki, K.M.; Zeredo, J.L.; et al. Periodontal and peri-implant wound healing following laser therapy. *Periodontol. 2000* **2015**, *68*, 217–269. [CrossRef]
2. Gholami, L.; Shahabi, S.; Jazaeri, M.; Hadilou, M.; Fekrazad, R. Clinical applications of antimicrobial photodynamic therapy in dentistry. *Front. Microbiol.* **2023**, *13*, 1020995. [CrossRef]
3. Takasaki, A.A.; Aoki, A.; Mizutani, K.; Schwarz, F.; Sculean, A.; Wang, C.Y.; Koshy, G.; Romanos, G.; Ishikawa, I.; Izumi, Y. Application of antimicrobial photodynamic therapy in periodontal and peri-implant diseases. *Periodontol. 2000* **2009**, *51*, 109–140. [CrossRef]
4. Chen, M.X.; Zhong, Y.J.; Dong, Q.Q.; Wong, H.M.; Wen, Y.F. Global, regional, and national burden of severe periodontitis, 1990-2019: An analysis of the Global Burden of Disease Study 2019. *J. Clin. Periodontol.* **2021**, *48*, 1165–1188. [CrossRef]
5. Janakiram, C.; Dye, B.A. A public health approach for prevention of periodontal disease. *Periodontol. 2000* **2020**, *84*, 202–214. [CrossRef]
6. Falcao, A.; Bullon, P. A review of the influence of periodontal treatment in systemic diseases. *Periodontol. 2000* **2019**, *79*, 117–128. [CrossRef]

7. Bui, F.Q.; Almeida-da-Silva, C.L.C.; Huynh, B.; Trinh, A.; Liu, J.; Woodward, J.; Asadi, H.; Ojcius, D.M. Association between periodontal pathogens and systemic disease. *Biomed. J.* **2019**, *42*, 27–35. [CrossRef]
8. Herrera, D.; Alonso, B.; Leon, R.; Roldan, S.; Sanz, M. Antimicrobial therapy in periodontitis: The use of systemic antimicrobials against the subgingival biofilm. *J. Clin. Periodontol.* **2008**, *35*, 45–66. [CrossRef]
9. Valen, H.; Scheie, A.A. Biofilms and their properties. *Eur. J. Oral Sci.* **2018**, *126*, 13–18. [CrossRef]
10. Quirynen, M.; Teughels, W.; van Steenberghe, D. Microbial shifts after subgingival debridement and formation of bacterial resistance when combined with local or systemic antimicrobials. *Oral Dis.* **2003**, *9*, 30–37. [CrossRef]
11. Walker, C.B. The acquisition of antibiotic resistance in the periodontal microflora. *Periodontol. 2000* **1996**, *10*, 79–88. [CrossRef]
12. Aoki, A.; Sasaki, K.M.; Watanabe, H.; Ishikawa, I. Lasers in nonsurgical periodontal therapy. *Periodontol. 2000* **2004**, *36*, 59–97. [CrossRef]
13. Schwarz, F.; Aoki, A.; Becker, J.; Sculean, A. Laser application in non-surgical periodontal therapy: A systematic review. *J. Clin. Periodontol.* **2008**, *35*, 29–44. [CrossRef]
14. Ebrahimi, P.; Hadilou, M.; Naserneysari, F.; Dolatabadi, A.; Tarzemany, R.; Vahed, N.; Nikniaz, L.; Fekrazad, R.; Gholami, L. Effect of photobiomodulation in secondary intention gingival wound healing-a systematic review and meta-analysis. *BMC Oral Health* **2021**, *21*, 258. [CrossRef]
15. Lopes, C.C.A.; Limirio, J.; Zanatta, L.S.A.; Simamoto, V.R.N.; Dechichi, P.; Limirio, A. Effectiveness of photobiomodulation therapy on human bone healing in dentistry: A systematic review. *Photobiomodul. Photomed. Laser Surg.* **2022**, *40*, 440–453. [CrossRef]
16. Mikami, R.; Mizutani, K.; Sasaki, Y.; Iwata, T.; Aoki, A. Patient-reported outcomes of laser-assisted pain control following non-surgical and surgical periodontal therapy: A systematic review and meta-analysis. *PLoS ONE* **2020**, *15*, e0238659. [CrossRef]
17. van Straten, D.; Mashayekhi, V.; de Bruijn, H.S.; Oliveira, S.; Robinson, D.J. Oncologic photodynamic therapy: Basic principles, current clinical status and future directions. *Cancers* **2017**, *9*, 19. [CrossRef]
18. Photodynamic therapy of subfoveal choroidal neovascularization in age-related macular degeneration with verteporfin: One-year results of 2 randomized clinical trials—DTAP report. Treatment of Age-related Macular Degeneration with Photodynamic Therapy (TAP) Study Group. Photodynamic therapy of subfoveal choroidal neovascularization in age-related macular degeneration with verteporfin: One-year results of 2 randomized clinical trials—TAP report. Treatment of age-related macular degeneration with photodynamic therapy (TAP) Study Group. *Arch. Ophthalmol.* **1999**, *117*, 1329–1345.
19. Kömerik, N.; Wilson, M.; Poole, S. The effect of photodynamic action on two virulence factors of gram-negative bacteria. *Photochem. Photobiol.* **2000**, *72*, 676–680. [CrossRef]
20. Braham, P.; Herron, C.; Street, C.; Darveau, R. Antimicrobial photodynamic therapy may promote periodontal healing through multiple mechanisms. *J. Periodontol.* **2009**, *80*, 1790–1798. [CrossRef]
21. Kousis, P.C.; Henderson, B.W.; Maier, P.G.; Gollnick, S.O. Photodynamic therapy enhancement of antitumor immunity is regulated by neutrophils. *Cancer Res.* **2007**, *67*, 10501–10510. [CrossRef]
22. Morgan, M.C.; Rashid, R.M. The effect of phototherapy on neutrophils. *Int. Immunopharmacol.* **2009**, *9*, 383–388. [CrossRef]
23. Tanaka, M.; Mroz, P.; Dai, T.; Huang, L.; Morimoto, Y.; Kinoshita, M.; Yoshihara, Y.; Nemoto, K.; Shinomiya, N.; Seki, S.; et al. Photodynamic therapy can induce a protective innate immune response against murine bacterial arthritis via neutrophil accumulation. *PLoS ONE* **2012**, *7*, e39823. [CrossRef]
24. Nakagawa, Y.; Homma, S.; Yamamoto, I.; Banno, M.; Nakazato, H.; Imanaga, H.; Yamamoto, N. In vivo and in vitro activation of macrophages with a cyanine photosensitizing dye, platonin. *Cancer Immunol. Immunother.* **1993**, *37*, 157–162. [CrossRef]
25. King, D.E.; Jiang, H.; Simkin, G.O.; Obochi, M.O.; Levy, J.G.; Hunt, D.W. Photodynamic alteration of the surface receptor expression pattern of murine splenic dendritic cells. *Scand. J. Immunol.* **1999**, *49*, 184–192. [CrossRef]
26. Preise, D.; Oren, R.; Glinert, I.; Kalchenko, V.; Jung, S.; Scherz, A.; Salomon, Y. Systemic antitumor protection by vascular-targeted photodynamic therapy involves cellular and humoral immunity. *Cancer Immunol. Immunother.* **2009**, *58*, 71–84. [CrossRef]
27. Séguier, S.; Souza, S.L.; Sverzut, A.C.; Simioni, A.R.; Primo, F.L.; Bodineau, A.; Correa, V.M.; Coulomb, B.; Tedesco, A.C. Impact of photodynamic therapy on inflammatory cells during human chronic periodontitis. *J. Photochem. Photobiol. B* **2010**, *101*, 348–354. [CrossRef]
28. Jiang, C.; Yang, W.; Wang, C.; Qin, W.; Ming, J.; Zhang, M.; Qian, H.; Jiao, T. Methylene Blue-Mediated Photodynamic Therapy Induces Macrophage Apoptosis via ROS and Reduces Bone Resorption in Periodontitis. *Oxid. Med. Cell Longev.* **2019**, *2019*, 1529520. [CrossRef]
29. da Cruz Andrade, P.V.; Euzebio Alves, V.T.; de Carvalho, V.F.; De Franco Rodrigues, M.; Pannuti, C.M.; Holzhausen, M.; De Micheli, G.; Conde, M.C. Photodynamic therapy decrease immune-inflammatory mediators levels during periodontal maintenance. *Lasers Med. Sci.* **2017**, *32*, 9–17. [CrossRef]
30. Salvi, G.E.; Stahli, A.; Schmidt, J.C.; Ramseier, C.A.; Sculean, A.; Walter, C. Adjunctive laser or antimicrobial photodynamic therapy to non-surgical mechanical instrumentation in patients with untreated periodontitis: A systematic review and meta-analysis. *J. Clin. Periodontol.* **2020**, *47*, 176–198. [CrossRef]
31. Dalvi, S.; Benedicenti, S.; Salagean, T.; Bordea, I.R.; Hanna, R. Effectiveness of antimicrobial photodynamic therapy in the treatment of periodontitis: A systematic review and meta-analysis of in vivo human randomized controlled clinical trials. *Pharmaceutics* **2021**, *13*, 836. [CrossRef]

32. Calciolari, E.; Ercal, P.; Dourou, M.; Akcali, A.; Tagliaferri, S.; Donos, N. The efficacy of adjunctive periodontal therapies during supportive periodontal care in patients with residual pockets. A systematic review and meta-analysis. *J. Periodontal. Res.* **2022**, *57*, 671–689. [CrossRef]
33. Moro, M.G.; de Carvalho, V.F.; Godoy-Miranda, B.A.; Kassa, C.T.; Horliana, A.; Prates, R.A. Efficacy of antimicrobial photodynamic therapy (aPDT) for nonsurgical treatment of periodontal disease: A systematic review. *Lasers Med. Sci.* **2021**, *36*, 1573–1590. [CrossRef]
34. Yoshida, A.; Sasaki, H.; Toyama, T.; Araki, M.; Fujioka, J.; Tsukiyama, K.; Hamada, N.; Yoshino, F. Antimicrobial effect of blue light using *Porphyromonas gingivalis* pigment. *Sci. Rep.* **2017**, *7*, 5225. [CrossRef]
35. Fontana, C.R.; Abernethy, A.D.; Som, S.; Ruggiero, K.; Doucette, S.; Marcantonio, R.C.; Boussios, C.I.; Kent, R.; Goodson, J.M.; Tanner, A.C.; et al. The antibacterial effect of photodynamic therapy in dental plaque-derived biofilms. *J. Periodontal Res.* **2009**, *44*, 751–759. [CrossRef]
36. Chui, C.; Hiratsuka, K.; Aoki, A.; Takeuchi, Y.; Abiko, Y.; Izumi, Y. Blue LED inhibits the growth of *Porphyromonas gingivalis* by suppressing the expression of genes associated with DNA replication and cell division. *Lasers Surg. Med.* **2012**, *44*, 856–864. [CrossRef]
37. Demidova, T.N.; Hamblin, M.R. Effect of cell-photosensitizer binding and cell density on microbial photoinactivation. *Antimicrob. Agents Chemother.* **2005**, *49*, 2329–2335. [CrossRef]
38. Sales, L.S.; Miranda, M.L.; de Oliveira, A.B.; Ferrisse, T.M.; Fontana, C.R.; Milward, M.; Brighenti, F.L. Effect of the technique of photodynamic therapy against the main microorganisms responsible for periodontitis: A systematic review of in-vitro studies. *Arch. Oral Biol.* **2022**, *138*, 105425. [CrossRef]
39. de Sousa, G.R.; Soares, L.O.; Soares, B.M.; de Carvalho Cruz, R.; Uliana Junior, P.; Santiago, T.; Farias, L.M.; Magalhaes, P.P.; Silveira, L.B.; Almeida Lopes, L.; et al. In vitro evaluation of physical and chemical parameters involved in aPDT of *Aggregatibacter actinomycetemcomitans*. *Lasers Med. Sci.* **2022**, *37*, 391–401. [CrossRef]
40. Akram, Z.; Shafqat, S.S.; Niaz, M.O.; Raza, A.; Naseem, M. Clinical efficacy of photodynamic therapy and laser irradiation as an adjunct to open flap debridement in the treatment of chronic periodontitis: A systematic review and meta-analysis. *Photodermatol. Photoimmunol. Photomed.* **2020**, *36*, 3–13. [CrossRef]
41. Bassir, S.H.; Moslemi, N.; Jamali, R.; Mashmouly, S.; Fekrazad, R.; Chiniforush, N.; Shamshiri, A.R.; Nowzari, H. Photoactivated disinfection using light-emitting diode as an adjunct in the management of chronic periodontitis: A pilot double-blind split-mouth randomized clinical trial. *J. Clin. Periodontol.* **2013**, *40*, 65–72. [CrossRef]
42. Pulikkotil, S.J.; Toh, C.G.; Mohandas, K.; Leong, K. Effect of photodynamic therapy adjunct to scaling and root planing in periodontitis patients: A randomized clinical trial. *Aust. Dent. J.* **2016**, *61*, 440–445. [CrossRef] [PubMed]
43. Husejnagic, S.; Lettner, S.; Laky, M.; Georgopoulos, A.; Moritz, A.; Rausch-Fan, X. Photoactivated disinfection in periodontal treatment: A randomized controlled clinical split-mouth trial. *J. Periodontol.* **2019**, *90*, 1260–1269. [CrossRef] [PubMed]
44. Harmouche, L.; Courval, A.; Mathieu, A.; Petit, C.; Huck, O.; Severac, F.; Davideau, J.L. Impact of tooth-related factors on photodynamic therapy effectiveness during active periodontal therapy: A 6-months split-mouth randomized clinical trial. *Photodiagnosis Photodyn. Ther.* **2019**, *27*, 167–172. [CrossRef] [PubMed]
45. Mongardini, C.; Di Tanna, G.L.; Pilloni, A. Light-activated disinfection using a light-emitting diode lamp in the red spectrum: Clinical and microbiological short-term findings on periodontitis patients in maintenance. A randomized controlled split-mouth clinical trial. *Lasers Med. Sci.* **2014**, *29*, 1–8. [CrossRef] [PubMed]
46. Goh, E.X.; Tan, K.S.; Chan, Y.H.; Lim, L.P. Effects of root debridement and adjunctive photodynamic therapy in residual pockets of patients on supportive periodontal therapy: A randomized split-mouth trial. *Photodiagnosis. Photodyn. Ther.* **2017**, *18*, 342–348. [CrossRef]
47. Hormdee, D.; Rinsathorn, W.; Puasiri, S.; Jitprasertwong, P. Anti-early stage of bacterial recolonization effect of curcuma longa extract as photodynamic adjunctive treatment. *Int. J. Dent.* **2020**, *2020*, 8823708. [CrossRef]
48. Ivanaga, C.A.; Miessi, D.M.J.; Nuernberg, M.A.A.; Claudio, M.M.; Garcia, V.G.; Theodoro, L.H. Antimicrobial photodynamic therapy (aPDT) with curcumin and LED, as an enhancement to scaling and root planing in the treatment of residual pockets in diabetic patients: A randomized and controlled split-mouth clinical trial. *Photodiagnosis Photodyn. Ther.* **2019**, *27*, 388–395. [CrossRef]
49. Bashkatov, A.; Genina, E.; Kochubey, V.; Tuchin, V. Optical properties of human skin, subcutaneous and mucous tissues in the wavelength range from 400 to 2000 nm. *J. Phys. D Appl. Phys.* **2005**, *38*, 2543–2555. [CrossRef]
50. Ronay, V.; Buchalla, W.; Sahrmann, P.; Attin, T.; Schmidlin, P.R. In vitro evaluation of the oxidation efficacy of transgingival photodynamic therapy. *Acta Odontol. Scand.* **2013**, *71*, 1216–1220. [CrossRef]
51. Wenzler, J.S.; Bocher, S.; Frankenberger, R.; Braun, A. Feasibility of transgingival laser irradiation for antimicrobial photodynamic therapy. *Photodiagnosis Photodyn. Ther.* **2019**, *28*, 75–79. [CrossRef] [PubMed]
52. Sasaki, Y.; Hayashi, J.I.; Fujimura, T.; Iwamura, Y.; Yamamoto, G.; Nishida, E.; Ohno, T.; Okada, K.; Yamamoto, H.; Kikuchi, T.; et al. New irradiation method with indocyanine green-loaded nanospheres for inactivating periodontal pathogens. *Int. J. Mol. Sci.* **2017**, *18*, 154. [CrossRef] [PubMed]
53. Schär, D.; Ramseier, C.A.; Eick, S.; Mettraux, G.; Salvi, G.E.; Sculean, A. Transgingival photodynamic therapy (tg-aPDT) adjunctive to subgingival mechanical instrumentation in supportive periodontal therapy. A randomized controlled clinical study. *Photodiagnosis Photodyn. Ther.* **2020**, *32*, 101971. [CrossRef] [PubMed]

54. Ichinose-Tsuno, A.; Aoki, A.; Takeuchi, Y.; Kirikae, T.; Shimbo, T.; Lee, M.C.; Yoshino, F.; Maruoka, Y.; Itoh, T.; Ishikawa, I.; et al. Antimicrobial photodynamic therapy suppresses dental plaque formation in healthy adults: A randomized controlled clinical trial. *BMC Oral Health* **2014**, *14*, 152. [CrossRef] [PubMed]
55. Hasanain, F.A.; Nassar, H.M. Utilizing Light Cure Units: A Concise Narrative Review. *Polymers* **2021**, *13*, 1596. [CrossRef] [PubMed]
56. Kunz, D.; Wirth, J.; Sculean, A.; Eick, S. In- vitro-activity of additive application of hydrogen peroxide in antimicrobial photodynamic therapy using LED in the blue spectrum against bacteria and biofilm associated with periodontal disease. *Photodiagnosis Photodyn. Ther.* **2019**, *26*, 306–312. [CrossRef]
57. Yuan, L.; Wang, Y.; Zong, Y.; Dong, F.; Zhang, L.; Wang, G.; Dong, H.; Wang, Y. Response of genes related to iron and porphyrin transport in *Porphyromonas gingivalis* to blue light. *J. Photochem. Photobiol. B* **2023**, *241*, 112670. [CrossRef]
58. Chui, C.; Aoki, A.; Takeuchi, Y.; Sasaki, Y.; Hiratsuka, K.; Abiko, Y.; Izumi, Y. Antimicrobial effect of photodynamic therapy using high-power blue light-emitting diode and red-dye agent on *Porphyromonas gingivalis*. *J. Periodontal Res.* **2013**, *48*, 696–705. [CrossRef]
59. Uekubo, A.; Hiratsuka, K.; Aoki, A.; Takeuchi, Y.; Abiko, Y.; Izumi, Y. Effect of antimicrobial photodynamic therapy using rose bengal and blue light-emitting diode on *Porphyromonas gingivalis* in vitro: Influence of oxygen during treatment. *Laser Ther.* **2016**, *25*, 299–308. [CrossRef]
60. Wang, D.; Pan, H.; Yan, Y.; Zhang, F. Rose bengal-mediated photodynamic inactivation against periodontopathogens in vitro. *Photodiagnosis Photodyn. Ther.* **2021**, *34*, 102250. [CrossRef]
61. Najafi, S.; Khayamzadeh, M.; Paknejad, M.; Poursepanj, G.; Kharazi Fard, M.J.; Bahador, A. An in vitro comparison of antimicrobial effects of curcumin-based photodynamic therapy and chlorhexidine, on *Aggregatibacter actinomycetemcomitans*. *J. Lasers Med. Sci.* **2016**, *7*, 21–25. [CrossRef] [PubMed]
62. Mahdi, Z.; Habiboallh, G.; Mahbobeh, N.N.; Mina, Z.J.; Majid, Z.; Nooshin, A. Lethal effect of blue light-activated hydrogen peroxide, curcumin and erythrosine as potential oral photosensitizers on the viability of *Porphyromonas gingivalis* and *Fusobacterium nucleatum*. *Laser Ther.* **2015**, *24*, 103–111. [CrossRef] [PubMed]
63. Saitawee, D.; Teerakapong, A.; Morales, N.P.; Jitprasertwong, P.; Hormdee, D. Photodynamic therapy of Curcuma longa extract stimulated with blue light against *Aggregatibacter actinomycetemcomitans*. *Photodiagnosis Photodyn. Ther.* **2018**, *22*, 101–105. [CrossRef] [PubMed]
64. Song, Y.; Lin, J.; Zhang, Z.; Xu, B.; Bi, L. Antimicrobial effect of photodynamic therapy using sinoporphyrin sodium and 390-400 nm light-emitting diode on *Porphyromonas gingivalis* in vitro. *Lasers Med. Sci.* **2021**, *36*, 153–164. [CrossRef] [PubMed]
65. Yoshida, A.; Inaba, K.; Sasaki, H.; Hamada, N.; Yoshino, F. Impact on *Porphyromonas gingivalis* of antimicrobial photodynamic therapy with blue light and Rose Bengal in plaque-disclosing solution. *Photodiagnosis Photodyn. Ther.* **2021**, *36*, 102576. [CrossRef] [PubMed]
66. Soukos, N.S.; Som, S.; Abernethy, A.D.; Ruggiero, K.; Dunham, J.; Lee, C.; Doukas, A.G.; Goodson, J.M. Phototargeting oral black-pigmented bacteria. *Antimicrob. Agents Chemother.* **2005**, *49*, 1391–1396. [CrossRef]
67. Feuerstein, O.; Ginsburg, I.; Dayan, E.; Veler, D.; Weiss, E.I. Mechanism of visible light phototoxicity on *Porphyromonas gingivalis* and *Fusobacterium nucleatum*. *Photochem. Photobiol.* **2005**, *81*, 1186–1189. [CrossRef]
68. Feuerstein, O.; Persman, N.; Weiss, E.I. Phototoxic effect of visible light on *Porphyromonas gingivalis* and *Fusobacterium nucleatum*: An in vitro study. *Photochem. Photobiol.* **2004**, *80*, 412–415. [CrossRef]
69. Izzo, A.D.; Walsh, J.T. Light-induced modulation of *Porphyromonas gingivalis* growth. *J. Photochem. Photobiol. B* **2004**, *77*, 63–69. [CrossRef]
70. Lee, J.; Song, H.Y.; Ahn, S.H.; Song, W.; Seol, Y.J.; Lee, Y.M.; Koo, K.T. In vitro investigation of the antibacterial and anti-inflammatory effects of LED irradiation. *J. Periodontal Implant Sci.* **2022**, *53*, 110–119. [CrossRef]
71. Soukos, N.S.; Mulholland, S.E.; Socransky, S.S.; Doukas, A.G. Photodestruction of human dental plaque bacteria: Enhancement of the photodynamic effect by photomechanical waves in an oral biofilm model. *Lasers Surg. Med.* **2003**, *33*, 161–168. [CrossRef]
72. Enwemeka, C.S. Antimicrobial blue light: An emerging alternative to antibiotics. *Photomed. Laser Surg.* **2013**, *31*, 509–511. [CrossRef]
73. Enwemeka, C.S.; Bumah, V.V.; Castel, J.C.; Suess, S.L. Pulsed blue light, saliva and curcumin significantly inactivate human coronavirus. *J. Photochem. Photobiol. B* **2022**, *227*, 112378. [CrossRef]
74. Bumah, V.V.; Morrow, B.N.; Cortez, P.M.; Bowman, C.R.; Rojas, P.; Masson-Meyers, D.S.; Suprapto, J.; Tong, W.G.; Enwemeka, C.S. The importance of porphyrins in blue light suppression of Streptococcus agalactiae. *J. Photochem. Photobiol. B* **2020**, *212*, 111996. [CrossRef]
75. Ashkenazi, H.; Malik, Z.; Harth, Y.; Nitzan, Y. Eradication of Propionibacterium acnes by its endogenic porphyrins after illumination with high intensity blue light. *FEMS Immunol. Med. Microbiol.* **2003**, *35*, 17–24. [CrossRef]
76. Masson-Meyers, D.S.; Bumah, V.V.; Biener, G.; Raicu, V.; Enwemeka, C.S. The relative antimicrobial effect of blue 405 nm LED and blue 405 nm laser on methicillin-resistant Staphylococcus aureus in vitro. *Lasers Med. Sci.* **2015**, *30*, 2265–2271. [CrossRef]
77. Araújo, N.C.; Fontana, C.R.; Gerbi, M.E.; Bagnato, V.S. Overall-mouth disinfection by photodynamic therapy using curcumin. *Photomed. Laser Surg.* **2012**, *30*, 96–101. [CrossRef]
78. Ricci Donato, H.A.; Pratavieira, S.; Grecco, C.; Brugnera-Junior, A.; Bagnato, V.S.; Kurachi, C. Clinical comparison of two photosensitizers for oral cavity decontamination. *Photomed. Laser Surg.* **2017**, *35*, 105–110. [CrossRef]

79. Genina, E.A.; Titorenko, V.A.; Belikov, A.V.; Bashkatov, A.N.; Tuchin, V.V. Adjunctive dental therapy via tooth plaque reduction and gingivitis treatment by blue light-emitting diodes tooth brushing. *J. Biomed. Opt.* **2015**, *20*, 128004. [CrossRef]
80. Kitanaka, Y.; Takeuchi, Y.; Hiratsuka, K.; Aung, N.; Sakamaki, Y.; Nemoto, T.; Meinzer, W.; Izumi, Y.; Iwata, T.; Aoki, A. The effect of antimicrobial photodynamic therapy using yellow-green LED and rose bengal on *Porphyromonas gingivalis*. *Photodiagnosis Photodyn. Ther.* **2020**, *32*, 102033. [CrossRef]
81. Crous, A.; Abrahamse, H. The signalling effects of photobiomodulation on osteoblast proliferation, maturation and differentiation: A review. *Stem Cell Rev. Rep.* **2021**, *17*, 1570–1589. [CrossRef] [PubMed]
82. Batista, L.F.Z.; Kaina, B.; Meneghini, R.; Menck, C.F.M. How DNA lesions are turned into powerful killing structures: Insights from UV-induced apoptosis. *Mutat. Res.* **2009**, *681*, 197–208. [CrossRef] [PubMed]
83. de Gruijl, F.R. Photocarcinogenesis: UVA vs UVB. *Methods Enzymol.* **2000**, *319*, 359–366. [CrossRef] [PubMed]
84. Pfeifer, G.P.; You, Y.H.; Besaratinia, A. Mutations induced by ultraviolet light. *Mutat. Res.* **2005**, *571*, 19–31. [CrossRef] [PubMed]
85. Takada, A.; Matsushita, K.; Horioka, S.; Furuichi, Y.; Sumi, Y. Bactericidal effects of 310 nm ultraviolet light-emitting diode irradiation on oral bacteria. *BMC Oral Health* **2017**, *17*, 96. [CrossRef] [PubMed]
86. Aung, N.; Aoki, A.; Takeuchi, Y.; Hiratsuka, K.; Katagiri, S.; Kong, S.; Shujaa Addin, A.; Meinzer, W.; Sumi, Y.; Izumi, Y. The effects of ultraviolet light-emitting diodes with different wavelengths on periodontopathic bacteria in vitro. *Photobiomodul. Photomed. Laser Surg.* **2019**, *37*, 288–297. [CrossRef] [PubMed]
87. Patrizi, A.; Raone, B.; Ravaioli, G.M. Management of atopic dermatitis: Safety and efficacy of phototherapy. *Clin. Cosmet. Investig. Dermatol.* **2015**, *8*, 511–520. [CrossRef]
88. Shintani, Y.; Yasuda, Y.; Kobayashi, K.; Maeda, A.; Morita, A. Narrowband ultraviolet B radiation suppresses contact hypersensitivity. *Photodermatol. Photoimmunol. Photomed.* **2008**, *24*, 32–37. [CrossRef]
89. Lapolla, W.; Yentzer, B.A.; Bagel, J.; Halvorson, C.R.; Feldman, S.R. A review of phototherapy protocols for psoriasis treatment. *J. Am. Acad. Dermatol.* **2011**, *64*, 936–949. [CrossRef]
90. Schweintzger, N.; Gruber-Wackernagel, A.; Reginato, E.; Bambach, I.; Quehenberger, F.; Byrne, S.N.; Wolf, P. Levels and function of regulatory T cells in patients with polymorphic light eruption: Relation to photohardening. *Br. J. Dermatol.* **2015**, *173*, 519–526. [CrossRef]
91. Deng, J.; Lu, C.; Zhao, Q.; Chen, K.; Ma, S.; Li, Z. The Th17/Treg cell balance: Crosstalk among the immune system, bone and microbes in periodontitis. *J. Periodontal Res.* **2022**, *57*, 246–255. [CrossRef] [PubMed]
92. Ebersole, J.L.; Kirakodu, S.; Novak, M.J.; Stromberg, A.J.; Shen, S.; Orraca, L.; Gonzalez-Martinez, J.; Burgos, A.; Gonzalez, O.A. Cytokine gene expression profiles during initiation, progression and resolution of periodontitis. *J. Clin. Periodontol.* **2014**, *41*, 853–861. [CrossRef] [PubMed]
93. Greene, A.C.; Shehabeldin, M.; Gao, J.; Balmert, S.C.; Ratay, M.; Sfeir, C.; Little, S.R. Local induction of regulatory T cells prevents inflammatory bone loss in ligature-induced experimental periodontitis in mice. *Sci. Rep.* **2022**, *12*, 5032. [CrossRef]
94. Wang, L.; Wang, J.; Jin, Y.; Gao, H.; Lin, X. Oral administration of all-trans retinoic acid suppresses experimental periodontitis by modulating the Th17/Treg imbalance. *J. Periodontol.* **2014**, *85*, 740–750. [CrossRef] [PubMed]
95. Agrawal, A.; Shindell, E.; Jordan, F.; Baeva, L.; Pfefer, J.; Godar, D.E. UV radiation increases carcinogenic risks for oral tissues compared to skin. *Photochem. Photobiol.* **2013**, *89*, 1193–1198. [CrossRef]
96. Breger, J.; Baeva, L.; Agrawal, A.; Shindell, E.; Godar, D.E. UVB-induced inflammatory cytokine release, DNA damage and apoptosis of human oral compared with skin tissue equivalents. *Photochem. Photobiol.* **2013**, *89*, 665–670. [CrossRef]
97. Araújo, P.V.; Teixeira, K.I.; Lanza, L.D.; Cortes, M.E.; Poletto, L.T. In vitro lethal photosensitization of S. mutans using methylene blue and toluidine blue O as photosensitizers. *Acta Odontol. Latinoam.* **2009**, *22*, 93–97.
98. Lin, J.; Bi, L.J.; Zhang, Z.G.; Fu, Y.M.; Dong, T.T. Toluidine blue-mediated photodynamic therapy of oral wound infections in rats. *Lasers Med. Sci.* **2010**, *25*, 233–238. [CrossRef]
99. Haffajee, A.D.; Socransky, S.S. Microbial etiological agents of destructive periodontal diseases. *Periodontol. 2000* **1994**, *5*, 78–111. [CrossRef]
100. Hajishengallis, G.; Lamont, R.J. Polymicrobial communities in periodontal disease: Their quasi-organismal nature and dialogue with the host. *Periodontol. 2000* **2021**, *86*, 210–230. [CrossRef]
101. Arany, P.R. Craniofacial wound healing with photobiomodulation therapy: New insights and current challenges. *J. Dent. Res.* **2016**, *95*, 977–984. [CrossRef] [PubMed]
102. Dompe, C.; Moncrieff, L.; Matys, J.; Grzech-Lesniak, K.; Kocherova, I.; Bryja, A.; Bruska, M.; Dominiak, M.; Mozdziak, P.; Skiba, T.H.I.; et al. Photobiomodulation-underlying mechanism and clinical applications. *J. Clin. Med.* **2020**, *9*, 1724. [CrossRef] [PubMed]
103. Lesniewski, A.; Estrin, N.; Romanos, G.E. Comparing the use of diode lasers to light-emitting diode phototherapy in oral soft and hard tissue procedures: A literature review. *Photobiomodul. Photomed. Laser Surg.* **2022**, *40*, 522–531. [CrossRef] [PubMed]
104. Nikinmaa, S.; Alapulli, H.; Auvinen, P.; Vaara, M.; Rantala, J.; Kankuri, E.; Sorsa, T.; Meurman, J.; Patila, T. Dual-light photodynamic therapy administered daily provides a sustained antibacterial effect on biofilm and prevents Streptococcus mutans adaptation. *PLoS ONE* **2020**, *15*, e0232775. [CrossRef] [PubMed]
105. Street, C.N.; Pedigo, L.A.; Loebel, N.G. Energy dose parameters affect antimicrobial photodynamic therapy-mediated eradication of periopathogenic biofilm and planktonic cultures. *Photomed. Laser Surg.* **2010**, *28*, S61–S66. [CrossRef] [PubMed]

106. de Freitas, M.T.M.; Soares, T.T.; Aragao, M.G.B.; Lima, R.A.; Duarte, S.; Zanin, I.C.J. Effect of photodynamic antimicrobial chemotherapy on mono- and multi-species cariogenic biofilms: A literature review. *Photomed. Laser Surg.* **2017**, *35*, 239–245. [CrossRef]
107. Teixeira, A.H.; Pereira, E.S.; Rodrigues, L.K.; Saxena, D.; Duarte, S.; Zanin, I.C. Effect of photodynamic antimicrobial chemotherapy on in vitro and in situ biofilms. *Caries Res.* **2012**, *46*, 549–554. [CrossRef]
108. Petelin, M.; Perkic, K.; Seme, K.; Gaspirc, B. Effect of repeated adjunctive antimicrobial photodynamic therapy on subgingival periodontal pathogens in the treatment of chronic periodontitis. *Lasers Med. Sci.* **2015**, *30*, 1647–1656. [CrossRef]
109. Ramanauskaite, E.; Moraschini, V.; Machiulskiene, V.; Sculean, A. Clinical efficacy of single and multiple applications of antimicrobial photodynamic therapy in periodontal maintenance: A systematic review and network meta-analysis. *Photodiagnosis Photodyn. Ther.* **2021**, *36*, 102435. [CrossRef]

antibiotics

MDPI

Article

Temporary Root Canal Obturation with a Calcium Hydroxide-Based Dressing: A Randomized Controlled Clinical Trial

Johannes-Simon Wenzler [1,2,*], Wolfgang Falk [3], Roland Frankenberger [2] and Andreas Braun [1,2]

1 Department of Operative Dentistry, Periodontology and Preventive Dentistry, Rheinisch-Westfälische Technische Hochschule University Hospital, Pauwelsstrasse 30, 52074 Aachen, Germany; anbraun@ukaachen.de
2 Department of Operative Dentistry, Endodontics and Pediatric Dentistry, Campus Marburg, University Medical Center Giessen and Marburg, Georg-Voigt-Strasse 3, 35039 Marburg, Germany
3 Center for Oro-Dental Microbiology, Hamburger Chausse 25, 24220 Flintbek, Germany
* Correspondence: jwenzler@ukaachen.de; Tel.: +49-(0)241-8037452

Abstract: Successful bacterial inactivation or elimination is essential for successful outcomes in endodontics. This study investigated the efficacy of a calcium hydroxide paste ($Ca(OH)_2$) as a temporary medical dressing for 1 week after chemomechanical root canal treatment (CMRCT). Microbiological samples from 26 patients were collected after endodontic emergency treatment as follows: (1) removal of the provisional filling material; (2) CMRCT; (3) irrigation with sodium hypochlorite I (3%); (4) medicinal insertion of $Ca(OH)_2$; and (5) irrigation with sodium hypochlorite II (3%). A microbiological examination was carried out after the specimens had been taken from the root canals via saline and sterile paper points. CMRCT resulted in a significant reduction in total bacterial load (TBL) in the root canal ($p < 0.05$). Additional irrigation (3) resulted in a further significant reduction in TBL ($p < 0.05$). In contrast, $Ca(OH)_2$ medication did not prevent the bacterial load from returning to the previous level immediately after CMRCT, but did not increase above that level either ($p < 0.05$). However, the increase in TBL was significant ($p < 0.05$) in comparison with the disinfection groups (I/II). Administration of $Ca(OH)_2$ for 1 week shows that in combination with an additional disinfection procedure, an increase in TBL must be expected, but not above the level of conditions after CMRCT.

Keywords: calcium hydroxide; endodontics; bacterial reduction; medicaments; disinfection

1. Introduction

Infection of the pulp complex can have many causes. Caries, trauma, or periodontitis, for example, are all associated with colonization of the root canal system, primarily with bacteria and fungi from the oral microbiome. Ultimately, bacterial penetration in most cases unavoidably leads to inflammation of the pulp or the periapical complex. Treatment then involves initiating chemomechanical root canal treatment, including medicinal treatment [1]. This is because effective microbial elimination is essential as part of systematic endodontic therapy, as failures in root canal treatment can usually be attributed to insufficient microbial reduction. Root canal irrigation, as the first step, is an essential part of this stage of treatment, since mechanical preparation of the root canals alone does not usually lead to sufficient microbial elimination. In contrast, chemomechanical preparation with adjunctive disinfection methods can eliminate more than 92% of the microorganisms in the endodont [2,3]. In addition to conventional rinsing solutions such as sodium hypochlorite and other adjuvant disinfection methods such as ultrasonic or laser systems, medicinal pastes temporarily applied to the root canal also have an important role to play. Among other things, they are considered to have a germ-inactivating or germ-eliminating effect.

One of these, calcium hydroxide, a widely used medication in endodontics, promises a bactericidal effect due to its high pH value and the release of hydroxyl ions, with only slight limitations on its effectiveness [4–6]. In vitro and in vivo studies have been published that demonstrate the effect in relation to different end points. Above all, microbiological tests—depending on the number of medicinal inserts and different previous disinfection protocols—show that calcium hydroxide pastes may be very effective [7,8]. In this context, other studies have underlined the importance of previously used disinfection measures in relation to residual bacteria in the root canal [9,10] and the resulting outcome [11]. However, microbial elimination never seems to be fully achieved, since even supplementary medicinal dressings with calcium hydroxide can only achieve almost complete sterility in 97% of cases, so that absolute sterility cannot be guaranteed in any case to date [12,13].

Against this background, the present in vivo study aimed to test a calcium hydroxide paste preparation on an aqueous basis with a pH of >12.5 and a calcium hydroxide content of 45% for its effectiveness in the context of root canal treatment. The study investigated the calcium hydroxide paste as a temporary root canal dressing after prior chemomechanical preparation in relation to the bacterial count reduction it achieved over a period of 1 week, testing the hypothesis that applying the dressing for a period of 1 week further reduces, or at least maintains at that level, the bacterial count achieved after disinfection.

2. Results

Table 1 and Figure 1 present the results for the microbiological and molecular biological samples collected ($n = 26$). At the start of the study (Baseline), a median value of 1.63×10^5 CFU (min. 1.11×10^4 max. 1.49×10^7, IQR 1.60×10^5) was observed.

Table 1. Total bacterial load in CFU/mL in the study groups at the different time points for sample collection.

	Baseline	Root Canal Treatment	Disinfection I	Medication (7d)	Disinfection II
Mean	9.05×10^5	1.65×10^5	5.12×10^4	3.34×10^5	1.02×10^5
Standard Deviation	2.99×10^6	4.33×10^5	8.53×10^4	1.18×10^6	3.35×10^5
Median	1.63×10^5	5.26×10^4	2.78×10^4	6.76×10^4	1.68×10^4
Minimum	1.11×10^4	8.81×10^2	1.98×10^2	2.80×10^3	1.89×10^2
Maximum	1.49×10^7	2.26×10^6	4.18×10^5	6.07×10^6	1.71×10^6
Interquartile Range	1.60×10^5	1.14×10^5	3.05×10^4	7.78×10^4	2.99×10^4
n	26	26	26	26	26

However, a statistically significant difference was noted after chemomechanical root canal preparation (Root Canal Preparation), with a median value of 5.26×10^4 CFU (min. 8.81×10^2, max. 2.26×10^6, IQR 1.14×10^5; $p < 0.05$).

After subsequent disinfection in accordance with the study protocol, a further statistically significant reduction in bacteria was observed (Disinfection I), with a median value of 2.78×10^4 CFU (min. 1.98×10^2, max. 4.18×10^5, IQR 3.05×10^4; $p < 0.05$).

Temporary root canal obturation with a calcium hydroxide paste [Medication (7d)] showed a median value of 6.76×10^4 CFU (min. 2.80×10^3, max. 6.07×10^6, IQR 7.78×10^4) and did not prevent the bacterial count from rising back to the level immediately after chemomechanical treatment (Root Canal Treatment), although there was no increase above that level. However, there was a statistically significant difference from subsequent disinfection before temporary calcium hydroxide medication ($p < 0.05$).

The second disinfection measure (Disinfection II), with a median value of 1.68×10^4 CFU (min. 1.89×10^2, max. 1.71×10^6, IQR 2.99×10^4) before gutta-percha obturation, again led to a statistically significant reduction in the total bacterial load (TBL) in comparison with (Medication (7d)) ($p < 0.05$).

Individual bacterial species were detected in our study, but only inconsistently. After consultation with the microbiology laboratory, TBL was therefore used as the main parameter for analysis.

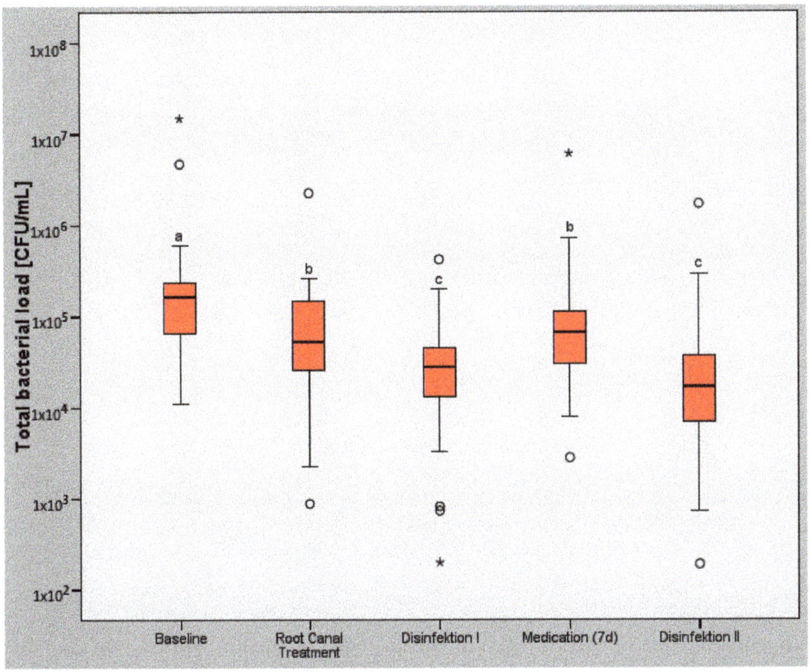

Figure 1. Box plot diagram for the total bacterial load (TBL) in the different groups; the same lowercase letters indicate no statistically significant differences ($p > 0.05$) between groups; mild outliners: ° and extreme outliners: *.

3. Materials and Methods

The present study was approved by the Ethics Committee (reference number 016/1749) and be carried out in full compliance with the current ethical principles (World Medical Association Declaration of Helsinki, version VI, 2002) at the Department of Operative Dentistry and Endodontology, University of Marburg, Germany. According to the diagnosis, clinical symptoms, and radiographic findings, teeth with suspected irreversible pulpitis were included, whereas teeth with suspected pulp necrosis were excluded. In addition, subjects had to be at least 18 years of age and had not received antibiotic treatment in the previous 6 months. Pregnancy, previous endodontic treatment of the tooth in concern, or a defect in the context of a periodontal-endodontic lesion were also considered as exclusion criteria. All 26 patients who fit the profile of the study according to the inclusion and exclusion criteria had previously given their consent to participate in the study.

3.1. *Treatment Procedure*

The initial endodontic emergency treatment of the study participants included, if necessary, restoration of the affected tooth with an adequate pre-endodontic composite restoration to ensure a suitable initial situation for further treatment. This then consisted of the isolation of the tooth with a rubber dam, the preparation of an access cavity, its irrigation with 3% sodium hypochlorite (5 mL), and an application of a calcium hydroxide paste (Calcicur; Voco GmbH, Cuxhaven, Germany). The tooth was then temporarily sealed using a foam pellet and a glass ionomer cement (Ketac Cem; 3M Espe, Seefeld, Germany). In the interests of patient welfare, no microbiological samples were taken at this point of the endodontic emergency treatment. One week after the emergency treatment, the patient returned for further treatment. In order to prevent bacterial contamination during the collection of the microbiological samples the tooth was again isolated with a rubber dam, which was then cleaned with a Lugol's iodine solution (5%) and subsequently inactivated with sodium thiosulfate (5%; Dr. Franz Köhler Chemie GmbH, Bensheim, Germany). To disinfect the tooth surface hydrogen peroxide solution (30%; Carl Roth GmbH, Karlsruhe, Germany) was used. From this point onwards, microbiological samples were taken five times during the further treatment sequence according to the study protocol from study arm I. The root canals were therefore flooded with sterile saline solution for one minute and the microbiological samples were collected at each of the following time points using sterile paper points (ISO 30; VDW Antaeos GmbH, Munich, Germany) [2]:

- First sampling (Baseline)—following the removal of the temporary filling material and calcium hydroxide paste (Figure 2).
- Second sampling (Root Canal Treatment)—after clinical screening and inclusion of appropriate teeth into the study, the included teeth were prepared via chemomechanical root canal preparation up to size 30.09 (ProTaper Gold; Dentsply Sirona GmbH, Bensheim, Germany) under irrigation with sodium hypochlorite (3%; 5 mL total, applied over the duration of the root canal preparation) and ethylenediaminetetraacetic acid (15%; 2 mL).
- Third sampling (Disinfection I)—following additional rinsing with sodium hypochlorite (5 mL) (within the same treatment session as the second sampling).
- Fourth sampling [Medication (7d)]–following the removal of the temporary filling and calcium hydroxide paste and only if the tooth has been free of symptoms for 1 week.
- Fifth sampling (Disinfection II)—following a final rinse with sodium hypochlorite (5 mL) (within the same treatment session as 4th sampling).

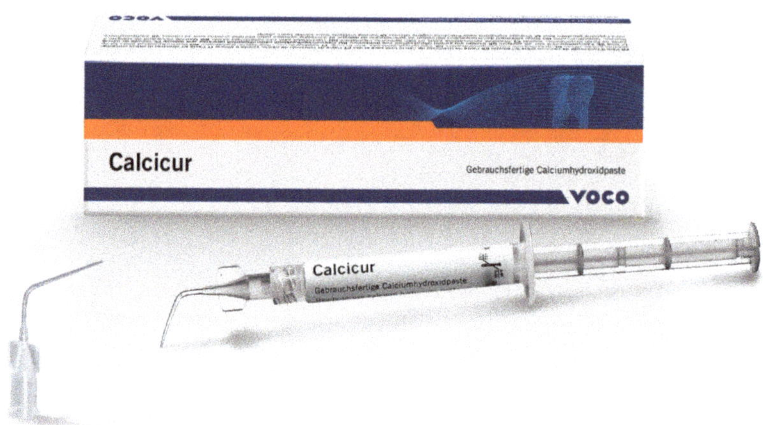

Figure 2. The calcium hydroxide paste used for temporary insertion into the root canal (Calcicur; Voco GmbH, Cuxhaven, Germany).

Subsequently, each root canal was obturated with a gutta-percha filling and a root canal sealer, as well as adhesive sealing of the access cavity.

All endodontic treatment procedures were performed according to a highly standardized protocol (study-internal standardization) to ensure comparability of the results, under the guidance of three dentists experienced in endodontics who were appointed as clinical investigators within the ethics approval as well as under the responsibility of the principal investigator responsible for the study.

3.2. Microbiological Analysis

An external laboratory (Oro-Dentale Mikrobiologie ODM, Kiel, Germany) performed the microbiological analysis of the previously collected samples. For this purpose, a quantitative real-time polymerase chain reaction (qPCR) was used and TBL was set as the main parameter for the analysis.

3.3. Preliminary Study

In a further study arm, bacterial colonization of the root canals was investigated in accordance with the procedure described above, but without the inclusion of a canal dressing with calcium hydroxide paste between treatments. After four patients had been treated, an interim evaluation was carried out, with results showing that the TBL, with a median of 8.95×10^4 CFU (min. 4.90×10^4, max. 1.45×10^5, IQR 3.68×10^4), indicated an increase in the bacterial count in comparison with the condition immediately after chemomechanical root canal preparation, with a median of 1.56×10^4 CFU (min. 1.36×10^4, max. 3.26×10^4, IQR 4.75×10^3)—an even greater divergence of the TBL in comparison with the Disinfection I group, with a median of 4.50×10^3 CFU (min. 1.80×10^3, max. 7.30×10^3, IQR 2.73×10^3). Since the omission of interim root canal dressing was not in accordance with standard clinical practice and resulted in an increase in the bacterial load in the root canal, this study arm was discontinued; in addition, it had no further clinical relevance to the working hypothesis of reducing, or at least maintaining, the bacterial reduction achieved prior to tooth closure by applying a calcium hydroxide paste.

3.4. Statistical Analysis

To estimate the number of patients required, a power analysis was performed prior to the study [14]. The data measured and documented in an Excel spreadsheet (Excel 2016, Microsoft Office Professional Plus 2016, Microsoft Corporation, Redmond, WA, USA) were statistically analyzed using IBM SPSS Statistics for Windows, version 26.0 (Version: 29.0.0.0 (241), IBM Corporation, Armonk, NY, USA). The Shapiro-Wilk test was used as a test for the normal distribution of the measured values. As not all data were normally distributed, values were analyzed with a nonparametric test (Kruskal–Wallis). A pairwise comparison was performed with the Mann-Whitney test. Comparisons within each study group were performed with nonparametric tests for paired samples (Friedman and Wilcoxon test). When multiple statistical tests were performed simultaneously on a single data set, Bonferroni correction of the critical p-value was applied. At $p < 0.05$, differences were deemed statistically significant. Boxplot diagrams display the minimum and maximum values (whiskers), as well as the median and first and third quartiles. Asterisks are used to identify outlier values, which are defined as values that are more than 1.5 to 3 times the interquartile range (IQR).

4. Discussion

The medicament most commonly used for temporary intracanal dressings is calcium hydroxide, which has a strong antibacterial effect due to its alkaline pH (12.5–12.8) and which is attributed to a resulting reduction of the intracanal bacterial load [6,15,16]. It is also assumed to have an influence on the outer root surface in terms of periodontal recovery [1,17,18], to be effective also against bacterial products such as lipopolysaccharides [19] and to control inflammatory exudates from the periapical area [20]. The results of the present study are largely consistent with previous findings regarding the basic assumptions about calcium hydroxide. In contrast to many previous studies, however, the results of the present study show that the total number of bacteria in the root canal increased significantly compared to the previous additional root canal irrigation (Disinfection I) despite the use of calcium hydroxide as a medicinal insert. Since the antibacterial efficacy of calcium hydroxide has previously been demonstrated in in vitro studies [4,21], the question remains as to how this renewed increase in bacterial colonization might be explained here. Considering that in vitro models of any kind can never reflect the complexity of actual in vivo situations, it must be taken into account that in vitro studies on antibacterial efficacy are often conducted using bacterial suspensions or artificial biofilms consisting of only a few bacterial species [22–25]. Therefore, other factors, such as additional protective properties of well-established biofilms may be of decisive importance here. These properties include a biofilm matrix, an altered growth rate of biofilm organisms, as well as other physiological changes [26,27]. Thus, most bacteria also showed increased resistance to alkaline challenges/stress when organized in the biofilms [26–28].

A possibly insufficient coating of the root canal wall with calcium hydroxide must also be considered here as a possible reason for the failure of the medicinal insert and the observed increase in TBL. A potential impairment of the coating of the root canal walls is the so-called vapor lock effect, which is considered a key limitation in the disinfection of root canals with rinsing solutions. This term refers to gas accumulations that usually occur in the lower third of the root canal (due to anatomical, physical, or chemical influences) and prevent the deeper penetration of rinsing solutions or medicinal pastes as well as their homogeneous distribution and thus also prevent their optimal effectiveness (e.g., due to lack of contact with the inner surfaces or dentinal tubules of the root canal) [29–31]. A recent study by Puleio et al. from 2023 investigated the vapor lock phenomenon during endodontic treatment using the CBCT technique and demonstrated its presence in almost all endodontic treatments, especially in the apical canal third [32]. With regard to the present study, it cannot be ruled out that the calcium hydroxide pastes used for medicinal inserts are also subjected to the phenomenon of the vapor-lock effect, thus impairing the sufficient coating of the entire surface of the root canal area. Furthermore, the ideal time that calcium hydroxide must be present in the root canal in order to comprehensively disinfect the canal system is not yet known. Nor is it known to which extent the type of bacteria as well as their location in the root canal influence the result [33]. Nevertheless, previous studies have already shown that up to 25% of bacteria can remain within the root canal, which is consistent with the results of the present [34–36]. As a result, remaining bacteria within the root canal could proliferate despite the use of a medicinal insert and thus cause a renewed increase in the TBL value at the time of the fourth sampling [Medication].

To date, only a few studies are available on the effectiveness of calcium hydroxide as a medicinal insert in vivo. Most of these studies primarily focused on treatment methodology and investigated single versus multiple-visit approaches in terms of disinfection and medicinal inserts. Nevertheless, interesting data concerning the question addressed in the present study can be obtained. An in vivo study on teeth with apical periodontitis by Vera et al. showed the importance of temporary medication of root canal–treated teeth from a histobacteriological point of view. In comparison with the so-called one-visit group (chemomechanical root canal treatment and obturation during the same appointment), the two-visit group (additional medication with calcium hydroxide paste for 1 week before obturation) showed an improved microbiological status. The study confirms the value of using calcium hydroxide paste for additional bacterial reduction and may provide an initial explanation for the rebound in total bacterial load observed in the present study as involving apparently surviving residual bacteria. The latter were found more frequently and in greater abundance in ramifications, isthmuses, and dentinal tubules—thus showing additional dependence of the results on the individual root canal anatomy [7]. In contrast to Vera et al. and the present study, Kvist et al. used bacterial cultures as evidence for bacterial reduction. The study also compared a one-visit group and a two-visit group with each other. The authors found that there were no statistically significant differences between the two study groups [8,15]. Unfortunately, closer comparison with the present study is not possible, as the study protocol used by Kvist et al. did not take samples between the individual disinfection procedures. However, from the results reported between the one-visit and two-visit groups, a presumed trend toward greater recolonization with bacteria can be interpreted—although, of course, in contrast to the present study, a potentially higher bacterial load due to the diagnosis of apical periodontitis should also be considered. The complexity of microbiological sampling and factors influencing it—

such as the anatomy of the root canal system in combination with temporary root canal pastes—must be particularly emphasized at this point [7,15,37]. The present study shows a high level of diagnostic accuracy in comparison with other approaches, free from the influence of calcium hydroxide. In contrast, it even shows that despite the temporary use of the substance, a transient increase in the total bacterial load is detectable—i.e., it does not lead to false-negative results [15,34]. Regardless of the initial situation and the treatment method, the situation in the current study shows a rather heterogeneous picture. Multiple application of calcium hydroxide is preferable to a single visit, and a storage time of 7–45 d appears to provide advantages [1,38,39]. It is possible that a longer calcium hydroxide storage period would also have shown a greater effect on bacterial reduction in this study. A new approach could possibly prove this.

In this context, however, attention should also be drawn to the actual disinfection protocols, i.e., to the use of disinfectant rinsing solutions. Even in the two disinfection groups (Disinfection I and II) in the present study, the significant impact of additional sodium hypochlorite irrigation on the total bacterial load can be seen. It is generally known that rinsing solutions such as sodium hypochlorite and chlorhexidine reduce the total bacterial load by up to 95% [14,16]. With regard to the efficacy of rinsing solutions alone, this is also confirmed by numerous publications and the resulting meta-analyses and systematic reviews of clinical trials, although it should be noted that the results are limited due to inconsistencies between articles and the lack of clinically relevant results, among other factors [40–42]. It is therefore not surprising that the two disinfection methods showed better disinfection efficacy in relation to the TBL, and a rebound can also be explained due to the limited effectiveness of the calcium hydroxide [43]. The studies by Siqueira et al. and Vianna et al. partly support the present results, but of course without the additional disinfection measures included in our protocol. Both studies investigate the combination of sodium hypochlorite and calcium hydroxide during root canal treatment, looking at the bactericidal effect. The results show that significant bacterial reduction can be expected after chemomechanical root canal preparation with sodium hypochlorite, but statistically significant differences were not found between root canal treatment and medicinal calcium hydroxide insertion (for 7 d). An additional significant bacteria-reducing effect by calcium hydroxide was therefore not confirmed [9,44]. This is also in line with our findings when comparing the study groups. Nonetheless, compared with the present data, it can be concluded that an additional disinfection measure during root canal treatment is essential in order to additionally reduce the TBL even after medication for 7 d.

Another important issue is that the increase in the TBL between the additional disinfection procedure [Disinfection I] and drug treatment after 7 days [Medication] could be due to presumed coronal leakage or to the skills of the operators in this study, in addition to the limited efficacy of calcium hydroxide. Coronal leakage, i.e., gaps between restorative materials and the cavity wall, caused by inadequate temporary fillings during or after endodontic therapy, may be responsible for bacterial recolonization and thus an increase in the TBL. In the present study protocol, the teeth were sealed using calcium hydroxide, a foam pellet, and a glass ionomer cement for 7 d. Recontamination can, of course, hardly be excluded. Several studies have shown advantages, especially through combinations of different materials or with self-adhesive products, but penetration tests using dye, bacteria, or glucose, for example, have never confirmed the complete impermeability of the different filling materials. In some cases, recontamination was detected within 48 h [45–47]. This could also explain a recurrent increase in the TBL between the (Disinfection I) and (Medication) groups in the present study, along with differences in the skills of the operators mentioned above. As different dentists were responsible for patient care in this study, the results are always dependent on the dentist concerned. For example, in the area of chemomechanical treatment, disinfection, calcium hydroxide application, calcium hydroxide removal, and temporary fillings mentioned above, variations can occur that naturally affect the results.

The use of the qPCR method and the omission of methods such as bacterial culture is also a factor worthy of discussion, which has already been mentioned. As in the study by Wenzler et al. [2], this study targeted the total load of bacteria previously defined as endodontically relevant, with the advantage that the qPCR method could be used specifically. Of course, as with the culture method, the opportunity was taken to evaluate the results openly—i.e., to detect bacteria not considered, or a variation in the spectrum. However, the PCR method has an advantage over the culture method for detecting difficult species, bacteria in a viable but nonculturable state, and species that have not yet been cultured. The problem with qPCR is certainly that dead cells or extracellular DNA can also lead to errors in the results due to amplification and detection [14,48,49]. However, this problem could be avoided in future studies by adding propidium monoazide (PMA), which can be used to separate dead from live bacteria [50]. In this study, qPCR demonstrated its advantages due to its sensitivity and specificity, as well as the time advantage in comparison with the culture method—also confirmed by the studies mentioned above. The qPCR method was considered suitable for this study.

With regard to the clinical significance of the present study, it should be noted that the observation period of the calcium hydroxide insert is limited to only one week and that no date on the overall (clinical and radiological) success of the endodontic treatments was included. Particularly with regard to the clinically relevant long-term success of endodontic treatments, it would be interesting to evaluate the follow-up of the patients treated in the scope of this study–possibly by means of a retrospective study. Nevertheless, within the limitations of this study, the results may support previous assumptions that successful root canal treatment depends on keeping the bacterial count below a threshold that the immune system can cope with. The long-term success of root canal therapy, however, cannot always be guaranteed due to a resurgence of bacteria in cases of immune suppression brought on by illness or, for example, aging [51]. Endodontic therapy should therefore always focus on the greatest possible reduction in bacteria. In addition to chemomechanical preparation and adjuvant disinfection protocols, which are generally and still considered to be the most important step in root canal disinfection, the introduction of intracanal medications such as calcium hydroxide is considered necessary in order to keep the bactericidal effects of the irrigation solutions constant and in addition to achieving maximum eradication of pathogens from the root canal [9,15,42,52].

The present study intentionally omitted an additional control group without calcium hydroxide placement (see Section 3.3 above, Preliminary Study). On the one hand, the calcium hydroxide dressing corresponds to the standard procedure for endodontic treatment in the clinic, and on the other hand, only the hypothesis of an additional reduction of the bacterial load in the root canal by the calcium hydroxide dressing over a period of 1 week was to be tested. This hypothesis was clearly rejected in the present study design. Further studies should follow, investigating the efficacy of other medicinal deposits that can effectively reduce the bacterial load in the temporarily closed root canal or at least maintain it at the level prior to closure.

5. Conclusions

Chemomechanical root canal treatment with a 3% sodium hypochlorite rinsing solution significantly reduced the bacterial count in the root canal, as did additional rinsing with sodium hypochlorite after the root canal treatment. Temporary administration of calcium hydroxide-based root canal medication for 7 d did not prevent the bacterial count from returning to the level immediately after chemomechanical treatment. The study shows that bacterial recolonization should be expected when a calcium hydroxide paste is used for temporary medicinal treatment of the root canal. However, the recolonization can be significantly reduced again by additional rinsing with sodium hypochlorite. Modern endodontics appears to be continuing to rely on the paradigm that disinfection with irrigation solutions rather than temporary root canal medication remains the essential element in endodontic therapy.

Author Contributions: Conceptualization, A.B.; methodology, J.-S.W. and A.B.; formal analysis, J.-S.W.; investigation, J.-S.W., microbiological analysis, W.F.; writing, original draft preparation, J.-S.W., R.F. and A.B.; writing, review and editing, J.-S.W., R.F. and A.B.; visualization, A.B.; project administration, A.B. and J.-S.W. All authors have read and agreed to the published version of the manuscript.

Funding: This research received no external funding.

Institutional Review Board Statement: The study was conducted in full accordance with established ethical principles (World Medical Association Declaration of Helsinki, version VI, 2002) and was approved by the local ethics committee (reference number 016/1749).

Informed Consent Statement: Informed consent was obtained from all patients included in the study.

Data Availability Statement: The data presented in this study are available upon request from the senior author, A.B.

Acknowledgments: The authors would like to thank the Voco GmbH company (Cuxhaven, Germany) for providing the calcium hydroxide paste used in the study.

Conflicts of Interest: The calcium hydroxide paste (Calcicur) was provided by the Voco GmbH company (Cuxhaven, Germany) to the senior author, A.B. To avoid any direct conflicts of interest, this author was not involved in practical aspects of the study.

References

1. Ordinola-Zapata, R.; Noblett, W.C.; Perez-Ron, A.; Ye, Z.; Vera, J. Present status and future directions of intracanal medicaments. *Int. Endod. J.* **2022**, *55* (Suppl. S3), 613–636. [CrossRef] [PubMed]
2. Wenzler, J.-S.; Falk, W.; Frankenberger, R.; Braun, A. Impact of Adjunctive Laser Irradiation on the Bacterial Load of Dental Root Canals: A Randomized Controlled Clinical Trial. *Antibiotics* **2021**, *10*, 1557. [CrossRef] [PubMed]
3. Shuping, G.B.; Orstavik, D.; Sigurdsson, A.; Trope, M. Reduction of intracanal bacteria using nickel-titanium rotary instrumentation and various medications. *J. Endod.* **2000**, *26*, 751–755. [CrossRef] [PubMed]
4. Georgopoulou, M.; Kontakiotis, E.; Nakou, M. In vitro evaluation of the effectiveness of calcium hydroxide and paramonochlorophenol on anaerobic bacteria from the root canal. *Endod. Dent. Traumatol.* **1993**, *9*, 249–253. [CrossRef] [PubMed]
5. Heithersay, G.S. Calcium hydroxide in the treatment of pulpless teeth with associated pathology. *J. Br. Endod. Soc.* **1975**, *8*, 74–93. [CrossRef] [PubMed]
6. Siqueira, J.F., Jr.; Lopes, H.P. Mechanisms of antimicrobial activity of calcium hydroxide: A critical review. *Int. Endod. J.* **1999**, *32*, 361–369. [CrossRef]
7. Vera, J.; Siqueira, J.F., Jr.; Ricucci, D.; Loghin, S.; Fernández, N.; Flores, B.; Cruz, A.G. One- versus two-visit endodontic treatment of teeth with apical periodontitis: A histobacteriologic study. *J. Endod.* **2012**, *38*, 1040–1052. [CrossRef]
8. Kvist, T.; Molander, A.; Dahlén, G.; Reit, C. Microbiological evaluation of one- and two-visit endodontic treatment of teeth with apical periodontitis: A randomized, clinical trial. *J. Endod.* **2004**, *30*, 572–576. [CrossRef]
9. Siqueira, J.F., Jr.; Guimarães-Pinto, T.; Rôças, I.N. Effects of chemomechanical preparation with 2.5% sodium hypochlorite and intracanal medication with calcium hydroxide on cultivable bacteria in infected root canals. *J. Endod.* **2007**, *33*, 800–805. [CrossRef]
10. Fabricius, L.; Dahlén, G.; Sundqvist, G.; Happonen, R.-P.; Möller, A.J.R. Influence of residual bacteria on periapical tissue healing after chemomechanical treatment and root filling of experimentally infected monkey teeth. *Eur. J. Oral Sci.* **2006**, *114*, 278–285. [CrossRef]
11. Sundqvist, G.; Figdor, D.; Persson, S.; Sjögren, U. Microbiologic analysis of teeth with failed endodontic treatment and the outcome of conservative re-treatment. *Oral Surg. Oral Med. Oral Pathol. Oral Radiol. Endod.* **1998**, *85*, 86–93. [CrossRef] [PubMed]
12. Sjögren, U.; Figdor, D.; Spångberg, L.; Sundqvist, G. The antimicrobial effect of calcium hydroxide as a short-term intracanal dressing. *Int. Endod. J.* **1991**, *24*, 119–125. [CrossRef] [PubMed]
13. Bystrom, A.; Claesson, R.; Sundqvist, G. The antibacterial effect of camphorated paramonochlorophenol, camphorated phenol and calcium hydroxide in the treatment of infected root canals. *Endod. Dent. Traumatol.* **1985**, *1*, 170–175. [CrossRef] [PubMed]
14. Blome, B.; Braun, A.; Sobarzo, V.; Jepsen, S. Molecular identification and quantification of bacteria from endodontic infections using real-time polymerase chain reaction. *Oral Microbiol. Immunol.* **2008**, *23*, 384–390. [CrossRef] [PubMed]
15. Sathorn, C.; Parashos, P.; Messer, H. Antibacterial efficacy of calcium hydroxide intracanal dressing: A systematic review and meta-analysis. *Int. Endod. J.* **2007**, *40*, 2–10. [CrossRef]
16. Koçkapan, C. *Curriculum Endodontie*; Quintessenz Publishing: Berlin, Germany, 2003.
17. Kawashima, N.; Wadachi, R.; Suda, H.; Yeng, T.; Parashos, P. Root canal medicaments. *Int. Dent. J.* **2009**, *59*, 5–11.
18. Duque, T.M.; Prado, M.; Herrera, D.R.; Gomes, B.P.F.A. Periodontal and endodontic infectious/inflammatory profile in primary periodontal lesions with secondary endodontic involvement after a calcium hydroxide-based intracanal medication. *Clin. Oral Investig.* **2019**, *23*, 53–63. [CrossRef]
19. Safavi, K.E.; Nichols, F.C. Alteration of biological properties of bacterial lipopolysaccharide by calcium hydroxide treatment. *J. Endod.* **1994**, *20*, 127–129. [CrossRef]

20. Chong, B.S.; Pitt Ford, T.R. The role of intracanal medication in root canal treatment. *Int. Endod. J.* **1992**, *25*, 97–106. [CrossRef]
21. Vatankhah, M.; Khosravi, K.; Zargar, N.; Shirvani, A.; Nekoofar, M.H.; Dianat, O. Antibacterial efficacy of antibiotic pastes versus calcium hydroxide intracanal dressing: A systematic review and meta-analysis of ex vivo studies. *J. Conserv. Dent.* **2022**, *25*, 463–480. [CrossRef]
22. Böcher, S.; Wenzler, J.-S.; Falk, W.; Braun, A. Comparison of different laser-based photochemical systems for periodontal treatment. *Photodiagnosis Photodyn. Ther.* **2019**, *27*, 433–439. [CrossRef] [PubMed]
23. Ratih, D.N.; Mulyawati, E.; Fajrianti, H. Antibacterial efficacy, calcium ion release, and pH using calcium hydroxide with three vehicles. *J. Conserv. Dent.* **2022**, *25*, 515–520. [CrossRef]
24. Lana, P.E.P.; Scelza, M.F.Z.; Silva, L.E.; Mattos-Guaraldi, A.L.d.; Hirata Júnior, R. Antimicrobial activity of calcium hydroxide pastes on Enterococcus faecalis cultivated in root canal systems. *Braz. Dent. J.* **2009**, *20*, 32–36. [CrossRef] [PubMed]
25. Wenzler, J.-S.; Krause, F.; Böcher, S.; Falk, W.; Birkenmaier, A.; Conrads, G.; Braun, A. Antimicrobial Impact of Different Air-Polishing Powders in a Subgingival Biofilm Model. *Antibiotics* **2021**, *10*, 1464. [CrossRef] [PubMed]
26. Stewart, P.S.; Costerton, J.W. Antibiotic resistance of bacteria in biofilms. *Lancet* **2001**, *358*, 135–138. [CrossRef]
27. Brändle, N.; Zehnder, M.; Weiger, R.; Waltimo, T. Impact of growth conditions on susceptibility of five microbial species to alkaline stress. *J. Endod.* **2008**, *34*, 579–582. [CrossRef]
28. Distel, J.W.; Hatton, J.F.; Gillespie, M.J. Biofilm formation in medicated root canals. *J. Endod.* **2002**, *28*, 689–693. [CrossRef]
29. Peters, O.A. Current challenges and concepts in the preparation of root canal systems: A review. *J. Endod.* **2004**, *30*, 559–567. [CrossRef]
30. Boutsioukis, C.; Kastrinakis, E.; Lambrianidis, T.; Verhaagen, B.; Versluis, M.; van der Sluis, L.W.M. Formation and removal of apical vapor lock during syringe irrigation: A combined experimental and Computational Fluid Dynamics approach. *Int. Endod. J.* **2014**, *47*, 191–201. [CrossRef]
31. Pesse, A.V.; Warrier, G.R.; Dhir, V.K. An experimental study of the gas entrapment process in closed-end microchannels. *Int. J. Heat Mass Transf.* **2005**, *48*, 5150–5165. [CrossRef]
32. Puleio, F.; Lizio, A.; Coppini, V.; Lo Giudice, R.; Lo Giudice, G. CBCT-Based Assessment of Vapor Lock Effects on Endodontic Disinfection. *Appl. Sci.* **2023**, *13*, 9542. [CrossRef]
33. Kurihara, H.; Kobayashi, Y.; Francisco, I.A.; Isoshima, O.; Nagai, A.; Murayama, Y. A microbiological and immunological study of endodontic-periodontic lesions. *J. Endod.* **1995**, *21*, 617–621. [CrossRef] [PubMed]
34. Reit, C.; Molander, A.; Dahlén, G. The diagnostic accuracy of microbiologic root canal sampling and the influence of antimicrobial dressings. *Endod. Dent. Traumatol.* **1999**, *15*, 278–283. [CrossRef] [PubMed]
35. Zehnder, M. Endodontic infection caused by localized aggressive periodontitis: A case report and bacteriologic evaluation. *Oral Surg. Oral Med. Oral Pathol. Oral Radiol. Endod.* **2001**, *92*, 440–445. [CrossRef]
36. Siqueira, J.F., Jr.; Rôças, I.N.; Lopes, H.P.; Magalhães, F.A.C.; de Uzeda, M. Elimination of Candida albicans infection of the radicular dentin by intracanal medications. *J. Endod.* **2003**, *29*, 501–504. [CrossRef]
37. Wu, M.-K.; Dummer, P.M.H.; Wesselink, P.R. Consequences of and strategies to deal with residual post-treatment root canal infection. *Int. Endod. J.* **2006**, *39*, 343–356. [CrossRef] [PubMed]
38. Nascimento, G.G.; Rabello, D.G.D.; Corazza, B.J.M.; Gomes, A.P.M.; Silva, E.G.; Martinho, F.C. Comparison of the effectiveness of single- and multiple-sessions disinfection protocols against endotoxins in root canal infections: Systematic review and meta-analysis. *Sci. Rep.* **2021**, *11*, 1226. [CrossRef]
39. Sharma, G.; Ahmed, H.M.A.; Zilm, P.S.; Rossi-Fedele, G. Antimicrobial properties of calcium hydroxide dressing when used for long-term application: A systematic review. *Aust. Endod. J.* **2018**, *44*, 60–65. [CrossRef]
40. Gonçalves, L.S.; Rodrigues, R.C.V.; Andrade Junior, C.V.; Soares, R.G.; Vettore, M.V. The Effect of Sodium Hypochlorite and Chlorhexidine as Irrigant Solutions for Root Canal Disinfection: A Systematic Review of Clinical Trials. *J. Endod.* **2016**, *42*, 527–532. [CrossRef]
41. Neelakantan, P.; Herrera, D.R.; Pecorari, V.G.A.; Gomes, B.P.F.A. Endotoxin levels after chemomechanical preparation of root canals with sodium hypochlorite or chlorhexidine: A systematic review of clinical trials and meta-analysis. *Int. Endod. J.* **2019**, *52*, 19–27. [CrossRef]
42. Ruksakiet, K.; Hanák, L.; Farkas, N.; Hegyi, P.; Sadaeng, W.; Czumbel, L.M.; Sang-Ngoen, T.; Garami, A.; Mikó, A.; Varga, G.; et al. Antimicrobial Efficacy of Chlorhexidine and Sodium Hypochlorite in Root Canal Disinfection: A Systematic Review and Meta-analysis of Randomized Controlled Trials. *J. Endod.* **2020**, *46*, 1032–1041.e7. [CrossRef] [PubMed]
43. Balto, K.A. Calcium hydroxide has limited effectiveness in eliminating bacteria from human root canal. *Evid. Based Dent.* **2007**, *8*, 15–16. [CrossRef]
44. Vianna, M.E.; Horz, H.-P.; Conrads, G.; Zaia, A.A.; Souza-Filho, F.J.; Gomes, B.P.F.A. Effect of root canal procedures on endotoxins and endodontic pathogens. *Oral Microbiol. Immunol.* **2007**, *22*, 411–418. [CrossRef] [PubMed]
45. Feliz Matos, L.; Rodriguez, I.D.L.S.; Gonzalez, M.L.R.; Pereyra, D.; Monzon Velez, E.R. Coronal microleakage of 3 temporary filling materials used for endodontic treatment: An in vitro study. *Gen. Dent.* **2013**, *61*, 52–55.
46. Wuersching, S.N.; Moser, L.; Obermeier, K.T.; Kollmuss, M. Microleakage of Restorative Materials Used for Temporization of Endodontic Access Cavities. *J. Clin. Med.* **2023**, *12*, 4762. [CrossRef] [PubMed]

47. Babu, N.S.V.; Bhanushali, P.V.; Bhanushali, N.V.; Patel, P. Comparative analysis of microleakage of temporary filling materials used for multivisit endodontic treatment sessions in primary teeth: An in vitro study. *Eur. Arch. Paediatr. Dent.* **2019**, *20*, 565–570. [CrossRef]
48. Brundin, M.; Figdor, D.; Roth, C.; Davies, J.K.; Sundqvist, G.; Sjögren, U. Persistence of dead-cell bacterial DNA in ex vivo root canals and influence of nucleases on DNA decay in vitro. *Oral Surg. Oral Med. Oral Pathol. Oral Radiol. Endod.* **2010**, *110*, 789–794. [CrossRef]
49. Rôças, I.N.; Siqueira, J.F., Jr. Identification of bacteria enduring endodontic treatment procedures by a combined reverse transcriptase-polymerase chain reaction and reverse-capture checkerboard approach. *J. Endod.* **2010**, *36*, 45–52. [CrossRef]
50. Meire, M.A.; van der Waal, S.V. A critical analysis of research methods and experimental models to study intracanal medicaments. *Int. Endod. J.* **2022**, *55* (Suppl. S2), 330–345. [CrossRef]
51. Arnold, M.; Arnold, W.; Attin, T.; Barbakow, F.; Benz, C.; Braun, A.; Bürkle, V.; Clauder, T.; Daubländer, M.; Edelhoff, D.; et al. *Endodontologie, 2., Überarbeitete und Erweiterte Auflage*; Georg Thieme Verlag KG: Stuttgart, Germany, 2008; ISBN 9783131902122.
52. Haapasalo, M.; Shen, Y.; Wang, Z.; Gao, Y. Irrigation in endodontics. *Br. Dent. J.* **2014**, *216*, 299–303. [CrossRef]

antibiotics MDPI

Review

The Interaction of Two Widely Used Endodontic Irrigants, Chlorhexidine and Sodium Hypochlorite, and Its Impact on the Disinfection Protocol during Root Canal Treatment

Dirk-Joachim Drews [1,2], Anh Duc Nguyen [2], Antje Diederich [2] and Christian Ralf Gernhardt [2,*]

[1] Private Dental Practice, 69469 Weinheim, Germany
[2] University Outpatient Clinic for Conservative Dentistry and Periodontology, Martin-Luther-University Halle-Wittenberg, 06112 Halle, Germany
[*] Correspondence: christian.gernhardt@uk-halle.de; Tel.: +49-345-557-3741

Abstract: In recent years, sodium hypochlorite and chlorhexidine digluconate have been the gold standard of irrigation solutions utilized within the disinfection protocol during root canal treatments. Nowadays, it is known that, during chemical disinfection of the root canal, consecutive application of sodium hypochlorite and chlorhexidine digluconate leads to the formation of an orange-brown precipitate. This precipitate is described as being chemically similar to para-chloroaniline, which is suspected to have cytotoxic and carcinogenic effects. Concerns also exist regarding its influence on the leakage of root canal fillings, coronal restorations, and tooth discoloration. The purpose of this article is to review the literature on the interaction of sodium hypochlorite and chlorhexidine digluconate on the tooth and its surrounding tissues, and to discuss the effect of the precipitate formed during root canal treatment. We further address options to avoid the formation of the precipitate and describe alternative irrigation solutions that should not interact with sodium hypochlorite or chlorhexidine digluconate.

Keywords: chlorhexidine digluconate; para-chloroaniline; precipitation; root canal irrigants; sodium hypochlorite

1. Introduction

Complete cleaning and disinfection of the root canal system are considered mandatory for long-term success in root canal treatment [1,2]. However, even after thorough mechanical cleaning, residual pulp tissue, bacteria, and dentin debris can remain in the root canal system [3,4]. Therefore, a variety of irrigating solutions are used in combination with the mechanical processing, such as sodium hypochlorite (NaOCl), chlorhexidine digluconate (CHX) [5], 17% ethylenediaminetetraacetic acid (EDTA), citric acid (CA), BioPure® MTAD® (Dentsply Tulsa Dental Specialties, Tulsa, OK, USA), and 37% phosphoric acid (PA) [6], as well as etidronate, alexidine (ALX), and Octenisept® (Schülke & Mayr, Norderstedt, Germany) [7]. Following internationally accepted quality guidelines, the main goals of irrigation are: eliminating microorganisms, flushing out debris, lubricating root canal instruments, and dissolving organic debris. Therefore, the used irrigation solution should preferably have disinfectant and organic-debris-dissolving properties, whilst not irritating the periradicular tissues [8]. For this purpose, sodium hypochlorite and chlorhexidine digluconate are widely recommended and well accepted in endodontics [9,10].

Unfortunately, endodontic irrigation solutions may interact chemically with each other during an alternating irrigation technique, potentially forming unwanted by-products, which may be toxic or cause allergic reactions [7]. Sodium hypochlorite and chlorhexidine are the best known and, at least in recent years, most frequently recommended irrigating solutions used for eliminating residual bacteria in chemo-mechanical root canal processing [5,6]. The undesirable adverse effects, after sodium hypochlorite and chlorhexidine interaction, of building precipitates are known, published and discussed controversially [11]. However, it is recommended that, until this precipitate is studied further, its formation should be avoided by removing the NaOCl before placing CHX into the canal [11]. Since 2006, the number of articles in PubMed concerning the interaction of NaOCl and CHX have grown significantly, and the topic was greatly debated [12–16]. Therefore, the aim of the present review is to summarize and discuss recently published papers focusing on the different outcomes regarding the interactions between sodium hypochlorite and chlorhexidine. Furthermore, based on the results of the review, the possible impact for the clinical disinfection protocol in endodontic therapy is summarized.

1.1. Sodium Hypochlorite (NaOCl)

Sodium hypochlorite (Figure 1) is the most used irrigating solution in endodontics, because its mechanism of action causes biosynthetic alterations in cellular metabolism and phospholipid destruction, the formation of chloramines that interfere in cellular metabolism, oxidative action with irreversible enzymatic inactivation of bacteria, and lipid and fatty acid degradation [17].

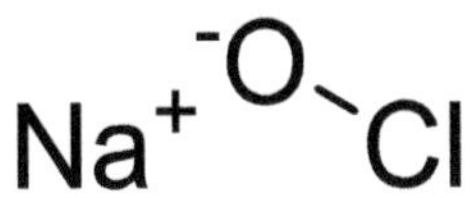

Figure 1. Structural formula of sodium hypochlorite.

Sodium hypochlorite (NaOCl) is the most common irrigant used in root canal treatments. NaOCl is an effective tissue solvent and antimicrobial agent. It is usually used in a concentration range from 0.5 to 8.25% [18–20]. Its germicidal ability is related to the formation of hypochlorous acid when in contact with organic debris. In high concentrations, NaOCl is toxic and can cause inflammation in the periapical tissues [21], whereas in low concentrations, it is ineffective against specific microorganisms. NaOCl is not a substantive antimicrobial agent; it tends to discolor and corrode surgical instruments; and it has a very unpleasant odor [11].

1.2. Chlorhexidine (CHX)

Chlorhexidine digluconate (CHX) is the gluconate salt form of chlorhexidine, a biguanide compound used as an antiseptic agent with topical antibacterial activity (Figure 2). Chlorhexidine digluconate is positively charged and reacts with the negatively charged microbial cell surface, thereby destroying the integrity of the cell membrane. Subsequently, chlorhexidine gluconate penetrates into the cell and causes a leakage of intracellular components, leading to cell death. Since gram-positive bacteria are more negatively charged, they are more sensitive to this agent [22].

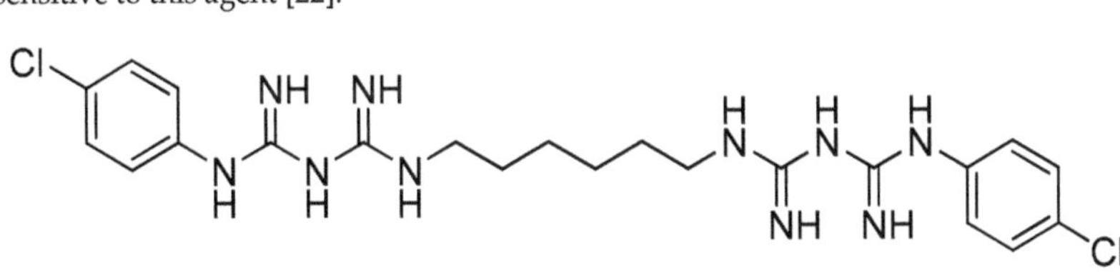

Figure 2. Structural formula of chlorhexidine gluconate.

Chlorhexidine digluconate (CHX) can be used as a complement to increase the antibacterial action of NaOCl solutions during root canal preparation. CHX shows similar antimicrobial effects to sodium hypochlorite [23,24] in vitro and possesses a lower toxicity [25,26]. A disadvantage compared to NaOCl is its lack of ability to dissolve vital and necrotic tissue [27].

Chlorhexidine digluconate is a broad-spectrum antibacterial agent with substantivity to tooth structures, i.e., it binds to the hydroxyapatite of the enamel and dentin or to anionic groups of glycoproteins, is slowly released and, due to the moderate concentration decrease, its antibacterial effects are prolonged for an extended period of time [28].

1.3. Proteolysis

When NaOCl and CHX are mixed, NaOCl dissociates into different ions (H^+, O^{2-}, and Cl^-). The chloride group then reacts with the chlorhexidine molecule in the guanine group (NH). This leads to the formation of chlorhexidine chloride (N^+ and Cl^-). In this reaction, the formation of an orange-brown precipitate is described. This precipitate contaminates the dentin and adheres to the canal walls [29]. Furthermore, CHX is a dicationic acid and has the ability to donate protons, whereas NaOCl is alkaline and can absorb protons from the dicationic acid. This proton exchange leads to the formation of a neutral and insoluble precipitate [11,30,31]. A color change due to the reaction can already be seen from a concentration of 0.023% NaOCl and the formation of the precipitate from 0.19% NaOCl by means of X-ray Photoelectron Spectroscopy (XPS), and the absolute amount by means of Time-of-Flight Secondary Ion Mass Spectrometry (ToF-SIMS) [11].

While the undesirable effects of the initial developing substances have been well studied and are classified as acceptable [5], the precipitate with regard to its ingredients and undesirable effects still gives rise to discussions [32,33].

Figure 3 demonstrates microtubes filled with 2% CHX mixed with different concentrations of NaOCl. The first microtube is a control sample with 2% CHX alone. From left to right, a color change, which becomes brighter as the concentration of NaOCl decreases, can be observed.

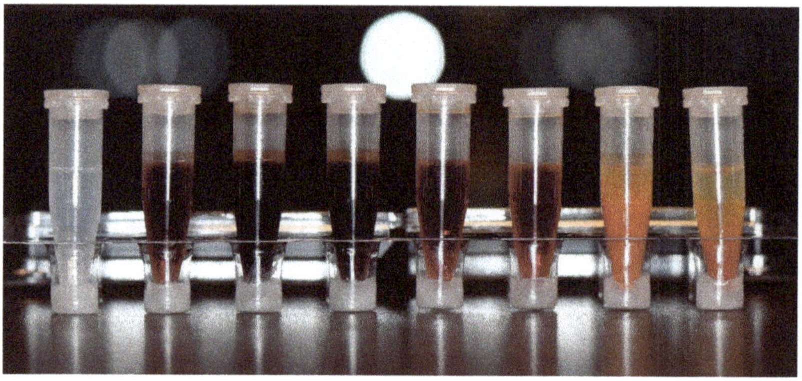

Figure 3. Microtubes containing 2% CHX mixed with different concentrations of NaOCl, to illustrate the precipitate formation. From left to right: (1) control sample with 0% NaOCl; (2) 0.5%; (3) 1%; (4) 1.5%; (5) 2.5%; (6) 3%; (7) 4%; (8) 5%.

2. Materials and Methods

An unlimited search in all fields of the PubMed database (https://pubmed.ncbi.nlm.nih.gov/, accessed on 4 October 2022) was carried out through the website of the National Center for Biotechnology Information (NCBI), utilizing the combination of the Medical Subject Headings (MeSH terms) "sodium hypochlorite" (NaOCl) AND "chlorhexidine" (CHX) and yielded 955 results from the years 1974–2022. Specifying the search term to "chlorhexidine AND sodium hypochlorite AND interaction", 64 publications remained from the original result. By individually reviewing the references and abstracts of these 64 publications the keywords "precipitate" and "para-chloroaniline" were regularly found in the keywords of the relevant articles (Figure 4).

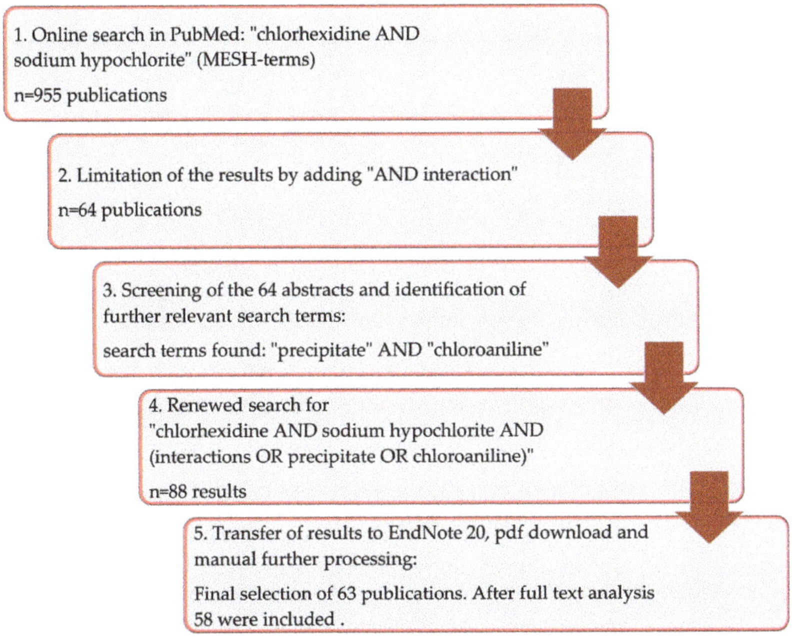

Figure 4. Graphical representation of the search strategy used in the present review.

Finally, the search term was further expanded to include "chlorhexidine AND sodium hypochlorite AND (interaction OR precipitate OR chloroaniline)", which resulted in a selection of 88 articles that included the manually determined references. The abstracts of all articles of the final online search result were evaluated and 25 articles that showed no relevance to the question were sorted out (Table 1, Figure 4).

Table 1. Inclusion and exclusion criteria.

Inclusion Criteria	Exclusion Criteria
- representing all search terms used	- articles without apparent relevance
- original articles, reviews, scientific short communications	- case reports, case series, editorials, case reviews

3. Results

The 63 papers included in this review are listed in Table 2, where the title and objective were summarized. These publications were read in full and evaluated.

58 publications (49 studies, 8 reviews, 1 short communication) were relevant to the topic; another 5 were excluded after reading the full texts. Sources to which the research publications referred were included if they were relevant to the topic, even if the date of their publication was before 1994.

Table 2. Included papers of the review.

Author	Title	Study Aim	Type
Bueso et al., 2022 [34]	Comparative evaluation of intermediate solutions in prevention of brown precipitate formed from sodium hypochlorite and chlorhexidine gluconate	To evaluate intermediate treatments between sodium hypochlorite and chlorhexidine gluconate irrigations for the prevention of a toxic brown precipitate in root canal therapy.	Laboratory study

Table 2. *Cont.*

Author	Title	Study Aim	Type
Jeong et al., 2021 [35]	Assessment of the cytotoxic effects and chemical composition of the insoluble precipitate formed from sodium hypochlorite and chlorhexidine gluconate	To investigate (1) the cytotoxic potential of the brown precipitate (BP) formed with sodium hypochlorite (NaOCl) and chlorhexidine gluconate (CHX), using both a small animal model of Caenorhabditis elegans (C. elegans) and cultured human gingival fibroblasts; (2) the chemical composition of BP using Time-of-Flight Secondary Ion Mass Spectrometry (ToF-SIMS).	Laboratory study
Czopik et al., 2021 [36]	Insight into the Reaction of Alexidine with Sodium Hypochlorite: A Potential Error in Endodontic Treatment	The aim of this study was to identify detected chemical compounds formed in the reaction of ALX and NaOCl with the ultra-high-performance liquid chromatography–mass spectrophotometry (UHPLC-MS) method and assess whether precipitates and PCA are formed in this reaction.	Laboratory study
Alberto et al., 2021 [37]	Does sodium thiosulphate avoid the formation of the brown-coloured precipitate as an intermediate irrigant between NaOCl and chlorhexidine?	This study evaluated the efficacy of sodium thiosulphate (ST) as an intermediate irrigant between sodium hypochlorite (NaOCl) and chlorhexidine (CHX) to avoid the formation of the brown-coloured precipitate.	Laboratory study
Khatib et al., 2020 [33]	Decoding the Perplexing Mystery of Para-Chloroaniline Formation: A Systematic Review	The purpose of this systematic review is to evaluate the relationship between PCA and brown precipitate.	Systematic review
Keles et al., 2020 [32]	Effect of various solutions on the removal of orange-brown precipitate formed by interaction of sodium hypochlorite and chlorhexidine with or without ultrasonic activation	The aim of this in vitro study was to investigate the possible interactions between photon-induced photoacoustic streaming (PIPS™)-activated oxidizing agents and 2% chlorhexidine digluconate.	Laboratory study
Buyukozer Ozkan et al., 2020 [38]	Proton Nuclear Magnetic Resonance Spectroscopy Analysis of Mixtures of Chlorhexidine with Different Oxidizing Agents Activated by Photon-Induced Photoacoustic Streaming for Root Canal Irrigation	The aim of the study was to assess the depth of sealer penetration into dentinal tubules following different final rinses and indirectly evaluate precipitation of irrigating solutions.	Laboratory study
Abusteit, 2020 [39]	Evaluation of resin sealer penetration of dentin following different final rinses for endodontic irrigation using confocal laser scanning microscopy	This study aimed to evaluate the characterization of chemical interaction of root canal irrigants on the surface of EndoSequence root repair materials using spectroscopy analysis.	Laboratory study
Abu Zeid et al., 2020 [40]	Morphological and chemical analysis of surface interaction of irrigant-endosequence root repair material	This study aimed to evaluate the characterization of chemical interaction of root canal irrigants on the surface of EndoSequence root repair materials using spectroscopy analysis.	Laboratory study
Thomas et al., 2019 [41]	Evaluation of the Antibacterial Efficiency of a Combination of 1% Alexidine and Sodium Hypochlorite on Enterococcus faecalis Biofilm Models: An In Vitro Study	The aim of the study was to assess the antibacterial efficiency of a combination of 1% alexidine (ALX) and 5.25% sodium hypochlorite (NaOCl) against E. faecalis biofilm using a confocal scanning electron microscopy.	Laboratory study
Siddique et al., 2019b [16]	Qualitative and quantitative analysis of precipitate formation following interaction of chlorhexidine with sodium hypochlorite, neem, and tulsi	This study aims to evaluate the precipitate formed on combination of different irrigants, weigh the amount of precipitate formed and to analyze the precipitate for PCA.	Laboratory study
Siddique et al., 2019a [42]	Quantitative analysis for detection of toxic elements in various irrigants, their combination (precipitate), and para-chloroaniline: An inductively coupled plasma mass spectrometry study	The aim of this study was to evaluate the precipitate formed on combination of different irrigants, weigh the amount of precipitate formed, and to analyze 35 different metal elements in each irrigant, precipitate formed as well as in PCA.	Laboratory study

Table 2. *Cont.*

Author	Title	Study Aim	Type
Žižka et al., 2018 [43]	Discoloration after Regenerative Endodontic Procedures: A Critical Review	This review presents a critical view on current knowledge of discoloration sources, its treatment and possible preventive modalities.	review
Ravinan-thanan et al., 2018 [44]	Cytotoxicity Evaluation of Combination Irrigant Regimens with MTAD on Two Different Cell Lines	The aim of this study was to evaluate the cytotoxicity of combination regimens on target and nontarget cell lines by trypan blue assay.	Laboratory study
Piperidou et al., 2018 [45]	Effects of Final Irrigation with SmearOFF on the Surface of Dentin Using Surface Analytical Methods	This study examined the chemical interaction of SmearOFF with sodium hypochlorite (NaOCl) on the dentin surface, specifically the formation of precipitate and/or parachloroaniline (PCA).	Laboratory study
Jain et al., 2018 [46]	Alexidine versus chlorhexidine for endodontic irrigation with sodium hypochlorite	The objective of this study was to chemically evaluate precipitate formation on irrigation by different concentrations of chlorhexidine (CHX) and alexidine (ALX) with sodium hypochlorite (NaOCl).	Laboratory study
Irmak et al., 2018 [47]	Nuclear magnetic resonance spectroscopy and infrared spectroscopy analysis of precipitate formed after mixing sodium hypochlorite and QMix 2 in 1	This study assessed whether para-chloroaniline (PCA) is formed after mixing NaOCl with Qmix.	Laboratory study
Gonzalez et al., 2018 [48]	Temperature changes in 2% chlorhexidine gluconate using two activation methods with different intensity levels	… the objective is to establish the influence of ultrasonic and sonic activation, with the use of different intensities, upon the temperature of chlorhexidine gluconate (CHX).	Laboratory study
Chhabra et al., 2018 [15]	Efficacy of various solutions in preventing orange-brown precipitate formed during alternate use of sodium hypochlorite and chlorhexidine: An In vitro study	The study evaluated the effectiveness of three Intermediate endodontic irrigating solutions in eliminating the residual sodium hypochlorite (NaOCl).	Laboratory study
Campbell et al., 2018 [49]	Antiseptics Commonly Used in Total Joint Arthroplasty Interact and May Form Toxic Products	Our clinical experience is that chlorhexidine (CHX) and Dakin's solution (NaOCl) interact and form a precipitate. The purpose of this study is to determine whether this reaction could be replicated in a laboratory setting, and to determine if other commonly used antiseptics also visibly react when mixed.	Laboratory study
Wright et al., 2017 [7]	Alkaline Sodium Hypochlorite Irrigant and Its Chemical Interactions	Of particular interest is the interaction between sodium hypochlorite and the chelators EDTA, citric acid and etidronate and between sodium hypochlorite and the antimicrobials chlorhexidine, alexidine, MTAD and octenisept.	review
Thaha et al., 2017 [50]	Interaction between Octenidine-based Solution and Sodium Hypochlorite: A Mass Spectroscopy, Proton Nuclear Magnetic Resonance, and Scanning Electron Microscopy-based Observational Study	The aim of this study was first to Identify the precipitate formed on the interaction between OCT and NaOCl and secondly to compare its effect on dentinal tubules with that of precipitate formed on combining chlorhexidine (CHX) and NaOCl.	Laboratory study
Surrender et al., 2017 [51]	Alexidine: A Safer and an Effective Root Canal Irrigant than Chlorhexidine	AIM: To compare antimicrobial activity of different concentrations of ALX with CHX individually and when combined with NaOCl against E. faecalis strains.	Laboratory study
Nocca et al., 2017 [14]	Chromographic Analysis and Cytotoxic Effects of Chlorhexidine and Sodium Hypochlorite Reaction Mixtures	This study aimed to investigate the stability of PCA in the presence of NaOCl and to examine the in vitro cytotoxic effects of CHX/NaOCl reaction mixtures.	Laboratory study
Guneser et al., 2017 [52]	Comparison of Conventional Syringe, CanalBrush, EndoActivator, Photon-Induced Photoacoustic Streaming, and Manual Instrumentation in Removing Orange-Brown Precipitate: An In Vitro Study	The aim of this In vitro study was to compare the various techniques for removing precipitate formed after irrigation with sodium hypochlorite (NaOCl) and chlorhexidine (CHX).	Laboratory study

Table 2. *Cont.*

Author	Title	Study Aim	Type
Vouzara et al., 2016 [53]	Combined and independent cytotoxicity of sodium hypochlorite, ethylenediaminetetraacetic acid and chlorhexidine	AIM: To evaluate the capacity of commonly used root canal irrigants to induce cytotoxic effects, when applied singly or in combination	Laboratory study
Patil et al., 2016 [54]	Determination of mutagenicity of the precipitate formed by sodium hypochlorite and chlorhexidine using the Ames test	The aim of this study was to determine the direct mutagenic potential of any precipitate formed by combining sodium hypochlorite (NaOCl) and chlorhexidine (CHX).	Laboratory study
Orhan et al., 2016 [55]	Does Para-chloroaniline Really Form after Mixing Sodium Hypochlorite and Chlorhexidine?	Purpose of this study was to determine whether PCA is formed through the reaction of mixing NaOCl and CHX.	Laboratory study
Mohammadi et al., 2015 [56]	Agonistic and Antagonistic Interactions between Chlorhexidine and Other Endodontic Agents: A Critical Review	The aim of this investigation was to review the agonistic and antagonistic interactions between chlorhexidine (CHX) and other irrigants and medicaments.	review
Metri et al., 2015 [57]	Comparative Evaluation of Two Final Irrigation Techniques for the Removal of Precipitate Formed by the Interaction between Sodium Hypochlorite and Chlorhexidine	AIM: To evaluate the effectiveness of two final irrigation techniques for the removal of precipitate formed by the interaction between sodium hypochlorite (NaOCl) and chlorhexidine (CHX).	Laboratory study
Magro et al., 2015 [58]	Effectiveness of several solutions to prevent the formation of precipitate due to the interaction between sodium hypochlorite and chlorhexidine and its effect on bond strength of an epoxy-based sealer	AIM: To evaluate the effectiveness of isopropyl alcohol, saline or distilled water to prevent the precipitate formed between sodium hypochlorite (NaOCl) and chlorhexidine (CHX) and its effect on the bond strength of an epoxy-based sealer in radicular dentine.	Laboratory study
Bernardi & Teixeira, 2015 [28]	The properties of chlorhexidine and undesired effects of its use in endodontics	The purpose of this article was to review the literature on the properties of chlorhexidine (CHX) and the adverse effects that may occur from its use in endodontics.	review
Arslan et al., 2015 [59]	Evaluation of orange-brown precipitate formed in root canals after irrigation with chlorhexidine and QMix and spectroscopic analysis of precipitates produced by a mixture of chlorhexidine/NaOCl and Qmix/NaOCl	AIM: To compare chlorhexidine and Qmix™ in terms of orange-brown precipitate generation in root canals and (ii) to analyse the precipitate produced by mixing chlorhexidine and Qmix(™) with NaOCl to determine whether para-chloroaniline was produced.	Laboratory study
Kolosowski et al., 2014 [60]	Qualitative analysis of precipitate formation on the surface and in the tubules of dentin irrigated with sodium hypochlorite and a final rinse of chlorhexidine or QMiX	The aim of this study was to qualitatively assess the formation of precipitate and PCA on the surface and in the tubules of dentin irrigated with NaOCl, followed either by EDTA, NaOCl, and CHX or by saline and QMiX.	Laboratory study
Homayouni et al., 2014 [61]	The Effect of Root Canal Irrigation with Combination of Sodium Hypo-chlorite and Chlorhexidine Gluconate on the Sealing Ability of Obturation Materials	The aim of this study was to evaluate the effect of the precipitate that was formed by combining Sodium Hypochlorite (NaOCl) and Chlorhexidine Gluconate (CHX) on the sealing ability of root canal obturation materials.	Laboratory study
Magro et al., 2014 [62]	Evaluation of the interaction between sodium hypochlorite and several formulations containing chlorhexidine and its effect on the radicular dentin—SEM and push-out bond strength analysis	The aim of the current study was to evaluate the presence of debris and smear layer after endodontic irrigation with different formulations of 2% chlorhexidine gluconate (CHX) and its effects on the push-out bond strength of an epoxy-based sealer on the radicular dentin.	Laboratory study
Cintra et al., 2014 [63]	The use of NaOCl in combination with CHX produces cytotoxic product	The aim of this study was to evaluate the tissue response to implanted polyethylene tubes filled with PPT-soaked fibrin sponge.	Laboratory study

Table 2. *Cont.*

Author	Title	Study Aim	Type
Arslan et al., 2014 [64]	Evaluation of effectiveness of various irrigating solutions on removal of calcium hydroxide mixed with 2% chlorhexidine gel and detection of orange-brown precipitate after removal	The aims of the present study were to evaluate the effect of various irrigating solutions on the removal of calcium hydroxide mixed with 2% chlorhexidine gel from an artificial groove created in a root canal and the generation of orange-brown precipitate in the remaining calcium hydroxide mixed with 2% chlorhexidine gel after irrigation with the various irrigating solutions.	Laboratory study
Souza et al., 2013 [65]	Evaluation of the colour change in enamel and dentine promoted by the interaction between 2% chlorhexidine and auxiliary chemical solutions	AIM: To evaluate the colour change in enamel and dentine, promoted by interaction of 2% chlorhexidine gluconate (CHX) with 5.25% sodium hypochlorite (NaOCl) and 17% ethylenediaminetetraacetic acid (EDTA).	Laboratory study
Shenoy et al., 2013 [66]	Assessment of precipitate formation on interaction of irrigants used in different combinations: an in vitro study	AIM: To evaluate the combination of various irrigants whether it forms the precipitate and also to quantify the amount of precipitate formed.	Laboratory study
Rossi-Fedele et al., 2013 [67]	Interaction between chlorhexidine-impregnated gutta-percha points and several chlorine-containing endodontic irrigating solutions	AIM: To evaluate if the immersion of chlorhexidine-impregnated gutta-percha points in chlorine-containing endodontic irrigants causes colour changes and precipitate formation.	Laboratory study
Prado et al., 2013 [68]	Interactions between irrigants commonly used in endodontic practice: a chemical analysis	The aim of this work was to characterize the by-products formed in the associations between the most commonly used irrigants in endodontic practice.	Laboratory study
Pasich et al., 2013 [69]	Efficacy of taurine haloamines and chlorhexidine against selected oral microbiome species	In this in vitro study we have compared antimicrobial activity of CHX with that of taurine chloramine (TauC1) and taurine bromamine (TauBr).	Laboratory study
Gupta et al., 2013 [70]	Evaluation of the sealing ability of two sealers after using chlorhexidine as a final irrigant: An in vitro study	The aim of this study was to evaluate the effect of the precipitate formed by using sodium hypochlorite and chlorhexidine as a root canal irrigant on the sealing ability of different root canal sealers.	Laboratory study
Gomes et al., 2013 [6]	Chlorhexidine in endodontics	The aim of this paper is to review CHX's general use in the medical field and in dentistry.	review
Vilanova et al., 2012 [71]	Effect of intracanal irrigants on the bond strength of epoxy resin-based and methacrylate resin-based sealers to root canal walls	AIM: To assess the bond strength of Epiphany and AH Plus sealers to root canal walls using a push-out test after use of several endodontic irrigants.	Laboratory study
Rossi-Fedele et al., 2012 [72]	Antagonistic interactions between sodium hypochlorite, chlorhexidine, EDTA, and citric acid	The aim of this investigation was to review the antagonistic interactions occurring when sodium hypochlorite (NaOCl), chlorhexidine (CHX), EDTA, and citric acid (CA) are used together during endodontic treatment.	review
Mortenson et al., 2012 [13]	The effect of using an alternative irrigant between sodium hypochlorite and chlorhexidine to prevent the formation of para-chloroaniline within the root canal system	AIM: To determine if the formation of para-chloroaniline (PCA) can be avoided by using an alternative irrigant following sodium hypochlorite but before chlorhexidine.	Laboratory study
Kim, 2012 [31]	Precipitate from a combination of sodium hypochlorite and chlorhexidine	... Chlorhexidine can form a precipitate when used in combination with NaOCl during intra-canal irrigation. What is the adverse effect of this precipitate and how can I reduce the chance of precipitation?	short communication

Table 2. *Cont.*

Author	Title	Study Aim	Type
Kim et al., 2012 [73]	Chemical interaction of alexidine and sodium hypochlorite	This study determined by electrospray ionization mass spectrometry (ESI-MS) and scanning electron microscopy (SEM) whether the chemical interaction between ALX and NaOCl results in PCA or precipitates.	Laboratory study
Gasic et al., 2012 [74]	Ultrastructural analysis of the root canal walls after simultaneous irrigation of different sodium hypochlorite concentration and 0.2% chlorhexidine gluconate	AIM: To determine whether sodium hypochlorite (NaOCl) with 0.2% chlorhexidine gluconate (CHX) leads to colour change and precipitate formation, and to ultrastructurally analyse the dentine surface after simultaneous irrigation with 0.5% NaOCl and 0.2% CHX.	Laboratory study
Prado et al., 2011 [75]	Effect of disinfectant solutions on the surface free energy and wettability of filling material	The aims of this study were to evaluate the surface free energy of GP and Res cones after disinfection procedures and to investigate the wettability of endodontic sealers in contact with these surfaces.	Laboratory study
Nowicki & Sem, 2011 [76]	An in vitro spectroscopic analysis to determine the chemical composition of the precipitate formed by mixing sodium hypochlorite and chlorhexidine	The purpose of this in vitro study was to determine the chemical composition of the precipitate formed by mixing sodium hypochlorite (NaOCl) and chlorhexidine (CHX) and the relative molecular weight of the components.	Laboratory study
de Assis et al., 2011 [77]	Evaluation of the interaction between endodontic sealers and dentin treated with different irrigant solutions	The aim of this study was to investigate the wettability of endodontic sealers in contact with dentin treated with 5.25% sodium hypochlorite (NaOCl) and 2% chlorhexidine (CHX) in the presence or absence of smear layer.	Laboratory study
Thomas & Sem, 2010 [78]	An in vitro spectroscopic analysis to determine whether para-chloroaniline is produced from mixing sodium hypochlorite and chlorhexidine	The purpose of this in vitro study was to determine whether para-chloroaniline (PCA) is formed through the reaction of mixing sodium hypochlorite (NaOCl) and chlorhexidine (CHX).	Laboratory study
Krishna-murthy & Sud-hakaran, 2010 [79]	Evaluation and prevention of the precipitate formed on interaction between sodium hypochlorite and chlorhexidine	The purpose of this study was (1) to evaluate maximum thickness the and chemical composition of the precipitate formed between sodium hypochlorite (NaOCl) and chlorhexidine (CHX) and (2) to evaluate effectiveness of absolute alcohol to remove residual NaOCl and thereby prevent the formation of the precipitate.	Laboratory study
Basrani et al., 2010 [80]	Determination of 4-chloroaniline and its derivatives formed in the interaction of sodium hypochlorite and chlorhexidine by using gas chromatography	The aim of this study was to further identify the precipitate by using gas chromatography-mass spectrometry (GC-MS).	Laboratory study
Akisue et al., 2010 [12]	Effect of the combination of sodium hypochlorite and chlorhexidine on dentinal permeability and scanning electron microscopy precipitate observation	This study compared the combined use of sodium hypochlorite (NaOCl) and chlorhexidine (CXH) with citric acid and CXH on dentinal permeability and precipitate formation.	Laboratory study
Mohammadi & Abbott, 2009 [81]	The properties and applications of chlorhexidine in endodontics	The purpose of this paper is to review the structure and mechanism of action of CHX, its antibacterial and antifungal activity, its effect on biofilm, its substantivity (residual antibacterial activity), its tissue solvent ability, its interaction with calcium hydroxide and sodium hypochlorite, its anticollagenolytic activity, its effect on coronal and apical leakage of bacteria, its toxicity and allergenicity and the modulating effect of dentine and root canal components on its antimicrobial activity.	review

Table 2. *Cont.*

Author	Title	Study Aim	Type
Basrani et al., 2009 [82]	Using diazotization to characterize the effect of heat or sodium hypochlorite on 2.0% chlorhexidine	The aim of the present study was to use a diazotization technique to confirm the presence of an aromatic amine (such as PCA) in the NaOCl/CHX precipitate and also in the 2.0% CHX at different temperatures (37 degrees C and 45 degrees C).	Laboratory study
Bui et al., 2008 [30]	Evaluation of the interaction between sodium hypochlorite and chlorhexidine gluconate and its effect on root dentin	The aim of this study was to evaluate the effect of irrigating root canals with a combination of NaOCl and CHX on root dentin and dentinal tubules ...	Laboratory study
Basrani et al., 2007 [11]	Interaction between sodium hypochlorite and chlorhexidine gluconate	The aim of this study was to determine the minimum concentration of NaOCl required to form a precipitate with 2.0% CHX.	Laboratory study
Zehnder, 2006 [5]	Root canal irrigants	In this review article, the specifics of the pulpal microenvironment and the resulting requirements for irrigating solutions are spelled out.	review

4. Discussion

The PubMed search found 63 publications, 58 of which were relevant. After full text analysis, 49 were studies that have been published since 2006 in the medical and especially in the dental-endodontic field, which have dealt with the interaction of NaOCl and CHX. Eight reviews with different focuses giving an overview of the state of knowledge at the date of publication were also selected. Furthermore, one article made recommendations on how to avoid formation of the precipitate, and thus was included.

4.1. Methodology

The 1998 study by Kuruvilla and Kamath [83] indicated that the alternating use of NaOCl and CHX reduces the microbial flora to a greater percentage (84.6%) than the use of NaOCl (69.4%) or CHX (70%) on its own. In order to optimize the tissue-dissolving properties of NaOCl and the antiseptic properties of the CHX against gram-positive germs, it was considered to use a combination of both irrigation solutions. However, by mixing the two solutions, for example through consecutive use in the root canal, a peach-colored to brown precipitate is formed [7], which is difficult to remove [10,13,16,17]. It is undisputed in the literature that the precipitate forms due to the acid–base reaction of NaOCl and CHX. The exact composition and, in particular, the question of whether the precipitate contains para-chloroaniline (PCA), motivated studies in the period from 2007 to 2021. Controversial views on the suitability of test methods for the analysis of the precipitate [13,78,80] and partly contradicting test results from the same test methods [47,59,76] leave doubts as to whether free PCA arises from the reaction of NaOCl and CHX [32]. However, recent studies emphasize the use of multiple non-destructive test methods and always examine 98% PCA as a comparison group, and they could not detect any free PCA in the precipitate [47,55]. In a review carried out by two independent authors on the basis of 13 included articles from different databases, Khatib et al. [33] concluded that the brown precipitate, which forms after mixing NaOCl and CHX, may contain a proportion of para-chloramide rather than free PCA and that PCA may be the by-product of the breakdown of highly concentrated CHX. It is also disputed whether PCA has mutagenic potential. While Gomes et al. [6] were citing publications from 1986 and 1995 according to which PCA was found to be mutagenic in microorganisms, Patil et al. [54] found no significant difference in the mutagenicity of the precipitate and the comparison group.

4.2. Toxicity

Regarding the toxicity of the precipitate, Cintra et al. [63] found a short-term increased toxicity compared to the starting substances, while Vouzara et al. [53] identified a predominantly antagonistic effect in the combination of NaOCl and CHX, indicating that the precipitate was less toxic than the starting substances. Surrender et al. [51] found the precipitate to be less toxic than either NaOCl or CHX alone. Furthermore, Jeong, Sarmast, Terlier, van der Hoeven, Holland and Parikh [35] all concluded that the precipitate has a toxic effect against human gingival fibroblasts, but highly concentrated NaOCl has an even greater cytotoxic effect. Nocca et al. [14] also observed a lower mortality of fibroblast cells to which the precipitate was applied than in those treated with the supernatant.

Marchesan et al. [84] evaluated the metals present in the precipitate of NaOCl and CHX by means of atomic absorption spectrophotometry and identified statistically significant proportions of copper (Cu), tin (Zn), iron (Fe), manganese (Mn), magnesium (Mg) and calcium (Ca). Siddique et al. [16] found selenium (Se) with inductively coupled plasma mass spectrometry. A discoloration of enamel and dentin was found by Souza et al. [65] on bovine anterior teeth that were placed in CHX gel and NaOCl consecutively. Therefore, it could be concluded that the combined use of NaOCl and CHX solution can cause dentin discoloration during endodontic treatment.

4.3. Recommended Irrigation Protocol

In order to prevent the formation of precipitates when using NaOCl and CHX, an irrigation protocol that includes intermediate rinses has frequently been recommended. For example, Zehnder [5] recommended rinsing the root canals exclusively with NaOCl during the mechanical preparation, which he ascribed to "unique tissue-dissolving properties". Before a final rinse with CHX recommended by him for chronic pulpitis and revisions, Zehnder [5] advised an intermediate rinse with EDTA or citric acid in order to prevent the formation of precipitates. It should be noted here that the root dentin can soften, if it is exposed to strong chelating agents, such as EDTA, for a long time [85]. Bueso et al. [34] used stereomicroscopic analysis to compare the effect of EDTA, distilled water and sodium thiosulfate (STS) as an intermediate rinse to prevent the formation of brown precipitates. In this context, 5% STS significantly reduced the intensity of brown precipitates, compared to no intermediate rinse. Alberto et al. [37] were able to demonstrate this effect ex vivo when CHX was added 10 min after the application of STS. Subsequent studies evaluated the endodontic irrigation regimen. The formation of precipitates was also demonstrated for the mixture of EDTA and NaOCl, but not for citric acid and CHX [72].

In addition, Mortenson et al. [13] found the least amounts of precipitate after intermediate flushing with 50% citric acid, compared to EDTA and saline. Intermediate rinsing with pure alcohol, distilled water, or saline solutions could also prevent or reduce the formation of precipitates [79]. However, a precipitate present in the root canal system represents a layer that occludes the dentinal tubules [12,30], is difficult to remove [29,79], and compromises the tightness of a root filling using AH 26 sealer (Dentsply Sirona, Konstanz, Germany) and gutta-percha proportionally to the amount of precipitate [61]. Whether the precipitate affects sealer adhesion has been controversially discussed: While Gupta et al. [70] came to the conclusion that the precipitate reduced the bonding capacity of an epoxy-based sealer (AH Plus®, Dentsply Sirona, Konstanz, Germany) significantly, it was subsequently shown that the adhesion of Resilon®-Epiphany SE obturation system (Pentron Clinical Technologies, Wallingford, CT, USA) was not affected by the precipitate. In addition, Magro et al. [58,62] found no correlation between the penetration depth of an epoxy sealer into the dentin and bond strength values between groups treated with or without CHX. However, the investigated CHX variants led to more precipitate in all root canal areas previously rinsed with NaOCl, though they did not reduce the bond strength of the sealer in the push-out test, which was traced back to the protocol for canal drying and covalent bonds between the sealer and the dentin surface [58,62].

Even by activating the rinsing solutions, the removal of the precipitate is only possible to a limited extent. However, it was found that activation of the chelating agents EDTA and citric acid, in particular using sonic (Eddy®, VDW, Munich, Germany) or ultrasound devices, is superior to syringe rinsing [32,52,57].

4.4. Alternative Irrigation Solutions

Furthermore, CHX alternatives were also considered and examined. The substitution of CHX by the herbal antimicrobial substances neem, tulsi, aloe vera, and garlic was not successful, as the amount of precipitate resulting from these substances in combination with NaOCl was a factor of 4–7.5 higher than that with CHX [42]. ALX, a substance from the biguanide family, similar to CHX, developed only a slightly yellowish color, but no precipitate formation with NaOCl [46]. In studies by Thomas et al. [41] the effectiveness of the combination of ALX and NaOCl against Enterococcus faecalis was not significantly higher than that of NaOCl alone. Nevertheless, the authors propagated that it can be used in 1% concentration as an alternative to CHX in the endodontic irrigation protocol if used for a sufficiently long time (>5 min). In contrast to the combination of NaOCl and CHX, Kim et al. [86] found no PCA in the mixture of NaOCl and ALX and considered it to be a CHX alternative because it is just as effective against all bacteria and fungi. Czopik [36] described a yellowish precipitate when mixing NaOCl and ALX, which could be identified as aliphatic amines by using the UHPLC-MS (ultra-high-performance liquid chromatography-mass spectrophotometry) method. Surender et al. [51] found a significantly higher effectiveness of NaOCl with ALX against Enterococcus faecalis than with the combination of NaOCl and CHX. Octenisept®, an octenidine-based preparation, which also contains 2% phenoxyethanol, led to a sparse, whitish deposits that partially closed the dentinal tubules and became transparent over time. Thaha et al. [50] saw potential for a combined application with NaOCl, but also a need for further investigations, for example, with regard to the effect on sealer adhesion to dentin. MTAD®, which contains 3% doxycycline, 4.25% CA, and 0.55% polysorbate, forms a green-yellow precipitate with NaOCl, the color of which changes to brown when exposed to light. Intermediate rinsing with ascorbic acid can prevent precipitation [7]. SmearOFF™ (Vista Apex, Raxine, WI, USA) and QMix® (Dentsply Sirona, Bensheim, Germany) are products that combine a biguanide and a chelator. After their application, the penetration depths of the sealer into the dentin were greater than those after sequential rinsing with 17% EDTA, saline solution, and CHX. While QMix® is used after saline or distilled water (2-phase), SmearOFF™ combines the intermediate rinse and the final rinse, which simplifies and shortens the rinsing protocol. According to the manufacturer, the use of SmearOFF™ after NaOCl in the root canal does not lead to the formation of precipitates; this is also indicated by the sealer penetration depths. Since the manufacturers have not disclosed the formulation of the preparations, further studies on effectiveness and interactions are required [39].

The possibility of exchanging NaOCl in the combination of NaOCl and CHX was only considered possible by Buyukozer et al. [38]. Chlorine dioxide (ClO_2) can be utilized as an alternative means of root canal irrigation instead of NaOCl, due to its antimicrobial activity, biocompatibility, and ability to dissolve organic tissue [87–89].

4.5. Clinical Impact on Endodontic Therapy

In 2023, the best possible cleaning and disinfection of the root canal system by means of chemomechanical preparation is still an indispensable prerequisite for the success of endodontic treatment [2,90,91].

The desirable properties of the various irrigation solutions are:

- Dissolution of necrotic and vital tissue;
- Effectiveness against bacteria;
- Effectiveness against fungi;
- Neutralization of endotoxins;
- Opening of the dentinal tubules;
- Removal of iatrogenic impurities;
- Economic efficiency;
- Practicality.

Unwanted properties are:

- Irritation of neighboring tissues;
- Cytotoxicity;
- Mutagenicity;
- Changes in the color of dentin or tooth enamel;
- Occlusion of the dentinal tubules;
- Undesirable interactions with other endodontic irrigating solutions and materials.

Since no irrigation solution is known that combines all the necessary properties and can be solely applied clinically, different solutions are used consecutively [92]. Certain combinations can interact with each other and result in undesirable effects or by-products. When in contact with each other in the root canal system, NaOCl and CHX interact in an acid–base reaction, forming an orange-brown precipitate, which has undesirable effects. The occlusion of the dentinal tubules [12,30] is indisputable and, depending on the sealer, can have a negative effect on its adhesive force or tightness [29,61,79]. The dyes of the precipitate can discolor the tooth substances, particularly the dentin.

Although it can be considered unlikely that the precipitate of NaOCl and CHX contains free para-chloroaniline, a substance that is suspected of being mutagenic, it is important to avoid precipitation in the root canal. After formation, the complete removal of the precipitate from the root canal system is difficult or impossible, even with advanced methods of activating irrigation solutions with sound, ultrasound, or laser pulses [32,57]. The safest method to avoid a precipitate from forming after use of NaOCl and CHX is to dispense only one of the two substances, which is the preferred option. Because of the sum of its properties, especially due to its ability to dissolve tissue, NaOCl is still the irrigation solution of choice during the mechanical preparation of the root canal. As long as the properties of potential alternatives, such as chlorine dioxide, have not been researched in more detail, it can still be regarded as the "gold standard" to use NaOCl exclusively in this phase [10].

In earlier studies, the combination of CHX and NaOCl was determined to have a better effect against Enterococcus faecalis and gram-positive germs compared to NaOCl alone [27,93–95]. Therefore, it was seen as an ideal complement to NaOCl. Some studies have since denied that the effectiveness of this combination against Enterococcus faecalis is better than that of NaOCl alone. However, this point is actually discussed controversially in the international literature [27,93–96].

A strict avoidance of the possible interaction between NaOCl and CHX in all its variants (including CHX solutions, CHX gels) is desired. According to the literature evaluated, this works best when an intermediate rinse with citric acid is utilized, as this removes the smear layer of the mechanical treatment without causing a precipitate with NaOCl or CHX and thereby triggering other complications.

In rinsing protocols that use NaOCl as the sole antimicrobial rinse, based on current knowledge, the final rinse to remove the smear layer should be carried out with citric acid or EDTA before the final use of NaOCl in an activated manner [10].

5. Conclusions

Since 2006, there has been a sharp increase in publications addressing the interactions between NaOCl and CHX. A total of 88 publications from the PubMed database were identified and evaluated. Of those, 58 publications were relevant to the topic. The results of the studies examined are often controversial, but certain aspects show a tendency over time.

The following findings relate to the endodontic irrigation protocol:

- The chemo-mechanical preparation of the root canal system is currently the gold standard;
- NaOCl should be used as the sole agent during mechanical reprocessing, due to its tissue-dissolving and antimicrobial properties;
- The smear layer can be removed with CA or EDTA after the mechanical preparation. NaOCl should not be mixed with CA or EDTA, since chelators neutralize the tissue-dissolving effect of NaOCl;
- The consecutive use of NaOCl and CHX is obsolete due to the precipitate that forms;

- If NaOCl and CHX (or CHX derivatives) are used in the same tooth, intermediate rinsing is required. Since CHX also forms a precipitate with EDTA, CA is recommended for this.

These recommendations are useful in clinical practice to effectively avoid the formation of the undesirable precipitate.

Author Contributions: Conceptualization, D.-J.D. and C.R.G.; methodology, D.-J.D. and C.R.G.; formal analysis, D.-J.D. and C.R.G.; investigation, D.-J.D.; writing—original draft preparation, D.-J.D., A.D.N., A.D. and C.R.G.; writing—review and editing, A.D.N. and C.R.G.; visualization, D.-J.D. and A.D.; supervision, C.R.G.; project administration, C.R.G. All authors have read and agreed to the published version of the manuscript.

Funding: This research received no external funding.

Institutional Review Board Statement: Not applicable.

Informed Consent Statement: Not applicable.

Data Availability Statement: Not applicable.

Conflicts of Interest: The authors declare no conflict of interest.

References

1. Kandaswamy, D.; Venkateshbabu, N. Root canal irrigants. *J Conserv. Dent.* **2010**, *13*, 256–264. [CrossRef]
2. Kojima, K.; Inamoto, K.; Nagamatsu, K.; Hara, A.; Nakata, K.; Morita, I.; Nakagaki, H.; Nakamura, H. Success rate of endodontic treatment of teeth with vital and nonvital pulps: A meta-analysis. *Oral Surg. Oral Med. Oral Pathol. Oral Radiol. Endod.* **2004**, *97*, 95–99. [CrossRef]
3. Abou-Rass, M.; Piccinino, M.V. The effectiveness of four clinical irrigation methods on the removal of root canal debris. *Oral Surg. Oral Med. Oral Pathol.* **1982**, *54*, 323–328. [CrossRef]
4. Bystrom, A.; Sundqvist, G. Bacteriologic evaluation of the efficacy of mechanical root canal instrumentation in endodontic therapy. *Scand. J. Dent. Res.* **1981**, *89*, 321–328. [CrossRef]
5. Zehnder, M. Root canal irrigants. *J. Endod.* **2006**, *32*, 389–398. [CrossRef]
6. Gomes, B.P.; Vianna, M.E.; Zaia, A.A.; Almeida, J.F.; Souza-Filho, F.J.; Ferraz, C.C. Chlorhexidine in endodontics. *Braz. Dent. J.* **2013**, *24*, 89–102. [CrossRef]
7. Wright, P.P.; Kahler, B.; Walsh, L.J. Alkaline Sodium Hypochlorite Irrigant and Its Chemical Interactions. *Materials* **2017**, *10*, 1147. [CrossRef]
8. European Society of Endodontology. Quality guidelines for endodontic treatment: Consensus report of the European Society of Endodontology. *Int. Endod. J.* **2006**, *39*, 921–930. [CrossRef]
9. Good, M.; El, K.I.; Hussey, D.L. Endodontic 'solutions' part 1: A literature review on the use of endodontic lubricants, irrigants and medicaments. *Dent. Update* **2012**, *39*, 239–240. [CrossRef]
10. Boutsioukis, C.; Arias-Moliz, M.T. Present status and future directions—Irrigants and irrigation methods. *Int. Endod. J.* **2022**, *55* (Suppl. S3), 588–612. [CrossRef] [PubMed]
11. Basrani, B.R.; Manek, S.; Sodhi, R.N.; Fillery, E.; Manzur, A. Interaction between sodium hypochlorite and chlorhexidine gluconate. *J. Endod.* **2007**, *33*, 966–969. [CrossRef] [PubMed]
12. Akisue, E.; Tomita, V.S.; Gavini, G.; Poli de Figueiredo, J.A. Effect of the combination of sodium hypochlorite and chlorhexidine on dentinal permeability and scanning electron microscopy precipitate observation. *J. Endod.* **2010**, *36*, 847–850. [CrossRef] [PubMed]
13. Mortenson, D.; Sadilek, M.; Flake, N.M.; Paranjpe, A.; Heling, I.; Johnson, J.D.; Cohenca, N. The effect of using an alternative irrigant between sodium hypochlorite and chlorhexidine to prevent the formation of para-chloroaniline within the root canal system. *Int. Endod. J.* **2012**, *45*, 878–882. [CrossRef] [PubMed]
14. Nocca, G.; Ahmed, H.M.A.; Martorana, G.E.; Callà, C.; Gambarini, G.; Rengo, S.; Spagnuolo, G. Chromographic Analysis and Cytotoxic Effects of Chlorhexidine and Sodium Hypochlorite Reaction Mixtures. *J. Endod.* **2017**, *43*, 1545–1552. [CrossRef] [PubMed]
15. Chhabra, N.; Gangaramani, S.; Singbal, K.P.; Desai, K.; Gupta, K. Efficacy of various solutions in preventing orange-brown precipitate formed during alternate use of sodium hypochlorite and chlorhexidine: An in vitro study. *J. Conserv. Dent.* **2018**, *21*, 428–432. [CrossRef]
16. Siddique, R.; Sureshbabu, N.M.; Somasundaram, J.; Jacob, B.; Selvam, D. Qualitative and quantitative analysis of precipitate formation following interaction of chlorhexidine with sodium hypochlorite, neem, and tulsi. *J. Conserv. Dent.* **2019**, *22*, 40–47. [CrossRef]
17. Estrela, C.; Estrela, C.R.; Barbin, E.L.; Spano, J.C.; Marchesan, M.A.; Pecora, J.D. Mechanism of action of sodium hypochlorite. *Braz. Dent. J.* **2002**, *13*, 113–117. [CrossRef]

18. Cullen, J.K.; Wealleans, J.A.; Kirkpatrick, T.C.; Yaccino, J.M. The effect of 8.25% sodium hypochlorite on dental pulp dissolution and dentin flexural strength and modulus. *J. Endod.* **2015**, *41*, 920–924. [CrossRef]
19. Hulsmann, M.; Hahn, W. Complications during root canal irrigation—Literature review and case reports. *Int. Endod. J.* **2000**, *33*, 186–193. [CrossRef]
20. Karkare, S.R.; Ahire, N.P.; Khedkar, S.U. Comparative evaluation of antimicrobial activity of hydroalcoholic extract of Aloe vera, garlic, and 5% sodium hypochlorite as root canal irrigants against Enterococcus faecalis: An in vitro study. *J. Indian Soc. Pedod. Prev. Dent.* **2015**, *33*, 274–278. [CrossRef]
21. Gernhardt, C.R.; Eppendorf, K.; Kozlowski, A.; Brandt, M. Toxicity of concentrated sodium hypochlorite used as an endodontic irrigant. *Int. Endod. J.* **2004**, *37*, 272–280. [CrossRef] [PubMed]
22. Ayhan, H.; Sultan, N.; Cirak, M.; Ruhi, M.Z.; Bodur, H. Antimicrobial effects of various endodontic irrigants on selected microorganisms. *Int. Endod. J.* **1999**, *32*, 99–102. [CrossRef]
23. Ercan, E.; Ozekinci, T.; Atakul, F.; Gul, K. Antibacterial activity of 2% chlorhexidine gluconate and 5.25% sodium hypochlorite in infected root canal: In vivo study. *J. Endod.* **2004**, *30*, 84–87. [CrossRef] [PubMed]
24. Rocas, I.N.; Siqueira, J.F., Jr. Comparison of the in vivo antimicrobial effectiveness of sodium hypochlorite and chlorhexidine used as root canal irrigants: A molecular microbiology study. *J. Endod.* **2011**, *37*, 143–150. [CrossRef]
25. Klein, U.; Kleier, D.J. Sodium hypochlorite accident in a pediatric patient. *Pediatr. Dent.* **2013**, *35*, 534–538.
26. Peters, O.A.; Paqué, F. Root canal preparation of maxillary molars with the self-adjusting file: A micro-computed tomography study. *J. Endod.* **2011**, *37*, 53–57. [CrossRef]
27. Gomes, B.P.; Martinho, F.C.; Vianna, M.E. Comparison of 2.5% sodium hypochlorite and 2% chlorhexidine gel on oral bacterial lipopolysaccharide reduction from primarily infected root canals. *J. Endod.* **2009**, *35*, 1350–1353. [CrossRef] [PubMed]
28. Bernardi, A.; Teixeira, C.S. The properties of chlorhexidine and undesired effects of its use in endodontics. *Quintessence Int.* **2015**, *46*, 575–582. [CrossRef]
29. Vivacqua-Gomes, N.; Ferraz, C.C.; Gomes, B.P.; Zaia, A.A.; Teixeira, F.B.; Souza-Filho, F.J. Influence of irrigants on the coronal microleakage of laterally condensed gutta-percha root fillings. *Int. Endod. J.* **2002**, *35*, 791–795. [CrossRef]
30. Bui, T.B.; Baumgartner, J.C.; Mitchell, J.C. Evaluation of the interaction between sodium hypochlorite and chlorhexidine gluconate and its effect on root dentin. *J. Endod.* **2008**, *34*, 181–185. [CrossRef]
31. Kim, J.W. Precipitate from a combination of sodium hypochlorite and chlorhexidine. *Restor. Dent. Endod.* **2012**, *37*, 185–186. [CrossRef]
32. Keles, A.; Ors, S.A.; Yilmaz, Z. Effect of various solutions on the removal of orange-brown precipitate formed by interaction of sodium hypochlorite and chlorhexidine with or without ultrasonic activationZ. *Niger. J. Clin. Pract.* **2020**, *23*, 381–385. [CrossRef] [PubMed]
33. Khatib, M.S.; Ameer, B.; Ajit Mannur, N.; Ramalingaiahsetty, A.M.; Peerzade, S.M.; Bambawale, A. Decoding the Perplexing Mystery of Para-Chloroaniline Formation: A Systematic Review. *J. Int. Soc. Prev. Community Dent.* **2020**, *10*, 142–147. [CrossRef]
34. Bueso, V.; Parikh, N.; Terlier, T.; Holland, J.N.; Sarmast, N.D.; Jeong, J.W. Comparative evaluation of intermediate solutions in prevention of brown precipitate formed from sodium hypochlorite and chlorhexidine gluconate. *Clin. Exp. Dent. Res.* **2022**, *8*, 1591–1597. [CrossRef]
35. Jeong, J.W.; Sarmast, N.D.; Terlier, T.; van der Hoeven, R.; Holland, J.N.; Parikh, N. Assessment of the cytotoxic effects and chemical composition of the insoluble precipitate formed from sodium hypochlorite and chlorhexidine gluconate. *Int. Endod. J.* **2021**, *54*, 1892–1901. [CrossRef]
36. Czopik, B.; Ciechomska, M.; Zarzecka, J.; Góra, M.; Woźniakiewicz, M. Insight into the Reaction of Alexidine with Sodium Hypochlorite: A Potential Error in Endodontic Treatment. *Molecules* **2021**, *26*, 1623. [CrossRef]
37. Alberto, A.P.L.; Oliveira, D.D.S.; Oliveira, H.E.; Maciel, A.C.C.; Belladonna, F.G.; Silva, E. Does sodium thiosulphate avoid the formation of the brown-coloured precipitate as an intermediate irrigant between NaOCl and chlorhexidine? *Aust. Endod. J.* **2022**, *48*, 72–76. [CrossRef]
38. Buyukozer Ozkan, H.; Terlemez, A.; Orhan, E.O. Proton Nuclear Magnetic Resonance Spectroscopy Analysis of Mixtures of Chlorhexidine with Different Oxidizing Agents Activated by Photon-Induced Photoacoustic Streaming for Root Canal Irrigation. *Photobiomodul. Photomed. Laser Surg.* **2020**, *38*, 374–379. [CrossRef] [PubMed]
39. Abusteit, O.E. Evaluation of resin sealer penetration of dentin following different final rinses for endodontic irrigation using confocal laser scanning microscopy. *Aust. Endod. J.* **2020**, *47*, 195–201. [CrossRef] [PubMed]
40. Abu Zeid, S.T.; Alamoudi, R.A.; Merdad, K. Morphological and chemical analysis of surface interaction of irrigant-endosequence root repair material. *J. Biomed. Mater. Res. B Appl. Biomater.* **2020**, *108*, 3383–3392. [CrossRef]
41. Thomas, A.R.; Mani, R.; Reddy, T.V.; Ravichandran, A.; Sivakumar, M.; Krishnakumar, S. Evaluation of the Antibacterial Efficiency of a Combination of 1% Alexidine and Sodium Hypochlorite on Enterococcus faecalis Biofilm Models: An In Vitro Study. *J. Contemp. Dent. Pract.* **2019**, *20*, 1090–1094. [CrossRef] [PubMed]
42. Siddique, R.; Nivedhitha, M.S.; Jacob, B. Quantitative analysis for detection of toxic elements in various irrigants, their combination (precipitate), and para-chloroaniline: An inductively coupled plasma mass spectrometry study. *J. Conserv. Dent.* **2019**, *22*, 344–350. [CrossRef] [PubMed]
43. Žižka, R.; Šedý, J.; Gregor, L.; Voborná, I. Discoloration after Regenerative Endodontic Procedures: A Critical Review. *Iran. Endod. J.* **2018**, *13*, 278–284. [CrossRef]

44. Ravinanthanan, M.; Hegde, M.N.; Shetty, V.; Kumari, S. Cytotoxicity Evaluation of Combination Irrigant Regimens with MTAD on Two Different Cell Lines. *Contemp. Clin. Dent.* **2018**, *9*, 255–259. [CrossRef]
45. Piperidou, M.; Sodhi, R.N.S.; Kolosowski, K.P.; Basrani, B.R. Effects of Final Irrigation with SmearOFF on the Surface of Dentin Using Surface Analytical Methods. *J. Endod.* **2018**, *44*, 1714–1719. [CrossRef] [PubMed]
46. Jain, K.; Agarwal, P.; Jain, S.; Seal, M.; Adlakha, T. Alexidine versus chlorhexidine for endodontic irrigation with sodium hypochlorite. *Eur. J. Dent.* **2018**, *12*, 398–402. [CrossRef]
47. Irmak, Ö.; Orhan, E.O.; Görgün, K.; Yaman, B.C. Nuclear magnetic resonance spectroscopy and infrared spectroscopy analysis of precipitate formed after mixing sodium hypochlorite and QMix 2in1. *PLoS ONE* **2018**, *13*, e0202081. [CrossRef]
48. Gonzalez, C.; Forner, L.; Llena, C.; Lozano, A. Temperature changes in 2% chlorhexidine gluconate using two activation methods with different intensity levels. *J. Clin. Exp. Dent.* **2018**, *10*, e458–e461. [CrossRef]
49. Campbell, S.T.; Goodnough, L.H.; Bennett, C.G.; Giori, N.J. Antiseptics Commonly Used in Total Joint Arthroplasty Interact and May Form Toxic Products. *J. Arthroplasty* **2018**, *33*, 844–846. [CrossRef]
50. Thaha, K.A.; Varma, R.L.; Nair, M.G.; Sam Joseph, V.G.; Krishnan, U. Interaction between Octenidine-based Solution and Sodium Hypochlorite: A Mass Spectroscopy, Proton Nuclear Magnetic Resonance, and Scanning Electron Microscopy-based Observational Study. *J. Endod.* **2017**, *43*, 135–140. [CrossRef]
51. Surender, L.R.; Shikha, A.; Swathi, A.; Manaswini, C.; Habeeb, A.; Prabha, S.S. Alexidine: A Safer and an Effective Root Canal Irrigant than Chlorhexidine. *J. Clin. Diagn. Res.* **2017**, *11*, ZC18–ZC21. [CrossRef]
52. Guneser, M.B.; Dincer, A.N.; Arslan, D. Comparison of Conventional Syringe, CanalBrush, EndoActivator, Photon-Induced Photoacoustic Streaming, and Manual Instrumentation in Removing Orange-Brown Precipitate: An In Vitro Study. *Photomed. Laser Surg.* **2017**, *35*, 311–316. [CrossRef] [PubMed]
53. Vouzara, T.; Koulaouzidou, E.; Ziouti, F.; Economides, N. Combined and independent cytotoxicity of sodium hypochlorite, ethylenediaminetetraacetic acid and chlorhexidine. *Int. Endod. J.* **2016**, *49*, 764–773. [CrossRef]
54. Patil, P.; Aminoshariae, A.; Harding, J.; Montagnese, T.A.; Mickel, A. Determination of mutagenicity of the precipitate formed by sodium hypochlorite and chlorhexidine using the Ames test. *Aust. Endod. J.* **2016**, *42*, 16–21. [CrossRef]
55. Orhan, E.O.; Irmak, Ö.; Hür, D.; Yaman, B.C.; Karabucak, B. Does Para-chloroaniline Really Form after Mixing Sodium Hypochlorite and Chlorhexidine? *J. Endod.* **2016**, *42*, 455–459. [CrossRef] [PubMed]
56. Mohammadi, Z.; Giardino, L.; Palazzi, F.; Asgary, S. Agonistic and Antagonistic Interactions between Chlorhexidine and Other Endodontic Agents: A Critical Review. *Iran. Endod. J.* **2015**, *10*, 1–5. [PubMed]
57. Metri, M.; Hegde, S.; Dinesh, K.; Indiresha, H.N.; Nagaraj, S.; Bhandi, S.H. Comparative Evaluation of Two Final Irrigation Techniques for the Removal of Precipitate Formed by the Interaction between Sodium Hypochlorite and Chlorhexidine. *J. Contemp. Dent. Pract.* **2015**, *16*, 850–853. [CrossRef] [PubMed]
58. Magro, M.G.; Kuga, M.C.; Aranda-Garcia, A.J.; Victorino, K.R.; Chávez-Andrade, G.M.; Faria, G.; Keine, K.C.; Só, M.V. Effectiveness of several solutions to prevent the formation of precipitate due to the interaction between sodium hypochlorite and chlorhexidine and its effect on bond strength of an epoxy-based sealer. *Int. Endod. J.* **2015**, *48*, 478–483. [CrossRef]
59. Arslan, H.; Uygun, A.D.; Keskin, A.; Karatas, E.; Seçkin, F.; Yıldırım, A. Evaluation of orange-brown precipitate formed in root canals after irrigation with chlorhexidine and QMix and spectroscopic analysis of precipitates produced by a mixture of chlorhexidine/NaOCl and QMix/NaOCl. *Int. Endod. J.* **2015**, *48*, 1199–1203. [CrossRef]
60. Kolosowski, K.P.; Sodhi, R.N.; Kishen, A.; Basrani, B.R. Qualitative analysis of precipitate formation on the surface and in the tubules of dentin irrigated with sodium hypochlorite and a final rinse of chlorhexidine or QMiX. *J. Endod.* **2014**, *40*, 2036–2040. [CrossRef]
61. Homayouni, H.; Majd, N.M.; Zohrehei, H.; Mosavari, B.; Adel, M.; Dajmar, R.; Homayouni, A. The Effect of Root Canal Irrigation with Combination of Sodium Hypo-chlorite and Chlorhexidine Gluconate on the Sealing Ability of Obturation Materials. *Open Dent. J.* **2014**, *8*, 184–187. [CrossRef] [PubMed]
62. Magro, G.M.; Kuga, M.C.; Regina Victorino, K.; Vazquez-Garcia, F.A.; Aranda-Garcia, A.J.; Faria-Junior, N.B.; Faria, G.; Luis Shinohara, A. Evaluation of the interaction between sodium hypochlorite and several formulations containing chlorhexidine and its effect on the radicular dentin—SEM and push-out bond strength analysis. *Microsc. Res. Tech.* **2014**, *77*, 17–22. [CrossRef] [PubMed]
63. Cintra, L.T.; Watanabe, S.; Samuel, R.O.; da Silva Facundo, A.C.; de Azevedo Queiroz, I.O.; Dezan-Júnior, E.; Gomes-Filho, J.E. The use of NaOCl in combination with CHX produces cytotoxic product. *Clin. Oral Investig.* **2014**, *18*, 935–940. [CrossRef] [PubMed]
64. Arslan, H.; Gok, T.; Saygili, G.; Altintop, H.; Akcay, M.; Capar, I.D. Evaluation of effectiveness of various irrigating solutions on removal of calcium hydroxide mixed with 2% chlorhexidine gel and detection of orange-brown precipitate after removal. *J. Endod.* **2014**, *40*, 1820–1823. [CrossRef]
65. Souza, M.; Cecchin, D.; Barbizam, J.V.; Almeida, J.F.; Zaia, A.A.; Gomes, B.P.; Ferraz, C.C. Evaluation of the colour change in enamel and dentine promoted by the interaction between 2% chlorhexidine and auxiliary chemical solutions. *Aust. Endod. J.* **2013**, *39*, 107–111. [CrossRef] [PubMed]
66. Shenoy, A.; Bolla, N.; Sarath, R.K.; Ram, C.H. Assessment of precipitate formation on interaction of irrigants used in different combinations: An in vitro study. *Indian J. Dent. Res.* **2013**, *24*, 451–455. [CrossRef]

67. Rossi-Fedele, G.; Doğramacı, E.J.; Steier, L.; de Figueiredo, J.A. Interaction between chlorhexidine-impregnated gutta-percha points and several chlorine-containing endodontic irrigating solutions. *Int. Endod. J.* **2013**, *46*, 675–680. [CrossRef]
68. Prado, M.; Santos Júnior, H.M.; Rezende, C.M.; Pinto, A.C.; Faria, R.B.; Simão, R.A.; Gomes, B.P. Interactions between irrigants commonly used in endodontic practice: A chemical analysis. *J. Endod.* **2013**, *39*, 505–510. [CrossRef] [PubMed]
69. Pasich, E.; Bialecka, A.; Marcinkiewicz, J. Efficacy of taurine haloamines and chlorhexidine against selected oral microbiome species. *Med. Dosw. Mikrobiol.* **2013**, *65*, 187–196.
70. Gupta, H.; Kandaswamy, D.; Manchanda, S.K.; Shourie, S. Evaluation of the sealing ability of two sealers after using chlorhexidine as a final irrigant: An in vitro study. *J. Conserv. Dent.* **2013**, *16*, 75–78. [CrossRef]
71. Vilanova, W.V.; Carvalho-Junior, J.R.; Alfredo, E.; Sousa-Neto, M.D.; Silva-Sousa, Y.T. Effect of intracanal irrigants on the bond strength of epoxy resin-based and methacrylate resin-based sealers to root canal walls. *Int. Endod. J.* **2012**, *45*, 42–48. [CrossRef]
72. Rossi-Fedele, G.; Doğramaci, E.J.; Guastalli, A.R.; Steier, L.; de Figueiredo, J.A. Antagonistic interactions between sodium hypochlorite, chlorhexidine, EDTA, and citric acid. *J. Endod.* **2012**, *38*, 426–431. [CrossRef]
73. Kim, H.S.; Zhu, Q.; Baek, S.H.; Jung, I.Y.; Son, W.J.; Chang, S.W.; Lee, W.; Gu, Y.; Lee, Y.; Hong, S.T.; et al. Chemical interaction of alexidine and sodium hypochlorite. *J. Endod.* **2012**, *38*, 112–116. [CrossRef]
74. Gasic, J.; Popovic, J.; Zivkovic, S.; Petrovic, A.; Barac, R.; Nikolic, M. Ultrastructural analysis of the root canal walls after simultaneous irrigation of different sodium hypochlorite concentration and 0.2% chlorhexidine gluconate. *Microsc. Res. Tech.* **2012**, *75*, 1099–1103. [CrossRef]
75. Prado, M.; de Assis, D.F.; Gomes, B.P.; Simão, R.A. Effect of disinfectant solutions on the surface free energy and wettability of filling material. *J. Endod.* **2011**, *37*, 980–982. [CrossRef] [PubMed]
76. Nowicki, J.B.; Sem, D.S. An in vitro spectroscopic analysis to determine the chemical composition of the precipitate formed by mixing sodium hypochlorite and chlorhexidine. *J. Endod.* **2011**, *37*, 983–988. [CrossRef] [PubMed]
77. De Assis, D.F.; Prado, M.; Simão, R.A. Evaluation of the interaction between endodontic sealers and dentin treated with different irrigant solutions. *J. Endod.* **2011**, *37*, 1550–1552. [CrossRef] [PubMed]
78. Thomas, J.E.; Sem, D.S. An in vitro spectroscopic analysis to determine whether para-chloroaniline is produced from mixing sodium hypochlorite and chlorhexidine. *J. Endod.* **2010**, *36*, 315–317. [CrossRef]
79. Krishnamurthy, S.; Sudhakaran, S. Evaluation and prevention of the precipitate formed on interaction between sodium hypochlorite and chlorhexidine. *J. Endod.* **2010**, *36*, 1154–1157. [CrossRef]
80. Basrani, B.R.; Manek, S.; Mathers, D.; Fillery, E.; Sodhi, R.N. Determination of 4-chloroaniline and its derivatives formed in the interaction of sodium hypochlorite and chlorhexidine by using gas chromatography. *J. Endod.* **2010**, *36*, 312–314. [CrossRef]
81. Mohammadi, Z.; Abbott, P.V. The properties and applications of chlorhexidine in endodontics. *Int. Endod. J.* **2009**, *42*, 288–302. [CrossRef]
82. Basrani, B.R.; Manek, S.; Fillery, E. Using diazotization to characterize the effect of heat or sodium hypochlorite on 2.0% chlorhexidine. *J. Endod.* **2009**, *35*, 1296–1299. [CrossRef]
83. Kuruvilla, J.R.; Kamath, M.P. Antimicrobial activity of 2.5% sodium hypochlorite and 0.2% chlorhexidine gluconate separately and combined, as endodontic irrigants. *J. Endod.* **1998**, *24*, 472–476. [CrossRef]
84. Marchesan, M.A.; Pasternak Junior, B.; Afonso, M.M.; Sousa-Neto, M.D.; Paschoalato, C. Chemical analysis of the flocculate formed by the association of sodium hypochlorite and chlorhexidine. *Oral Surg. Oral Med. Oral Pathol. Oral Radiol. Endod.* **2007**, *103*, e103–e105. [CrossRef] [PubMed]
85. Calt, S.; Serper, A. Time-dependent effects of EDTA on dentin structures. *J. Endod.* **2002**, *28*, 17–19. [CrossRef] [PubMed]
86. Kim, H.S.; Woo Chang, S.; Baek, S.H.; Han, S.H.; Lee, Y.; Zhu, Q.; Kum, K.Y. Antimicrobial effect of alexidine and chlorhexidine against Enterococcus faecalis infection. *Int. J. Oral Sci.* **2013**, *5*, 26–31. [CrossRef]
87. Cobankara, F.K.; Ozkan, H.B.; Terlemez, A. Comparison of organic tissue dissolution capacities of sodium hypochlorite and chlorine dioxide. *J. Endod.* **2010**, *36*, 272–274. [CrossRef] [PubMed]
88. Herczegh, A.; Ghidan, A.; Friedreich, D.; Gyurkovics, M.; Bendo, Z.; Lohinai, Z. Effectiveness of a high purity chlorine dioxide solution in eliminating intracanal Enterococcus faecalis biofilm. *Acta Microbiol. Immunol. Hung.* **2013**, *60*, 63–75. [CrossRef] [PubMed]
89. Nishikiori, R.; Nomura, Y.; Sawajiri, M.; Masuki, K.; Hirata, I.; Okazaki, M. Influence of chlorine dioxide on cell death and cell cycle of human gingival fibroblasts. *J. Dent.* **2008**, *36*, 993–998. [CrossRef]
90. Fleming, C.H.; Litaker, M.S.; Alley, L.W.; Eleazer, P.D. Comparison of classic endodontic techniques versus contemporary techniques on endodontic treatment success. *J. Endod.* **2010**, *36*, 414–418. [CrossRef]
91. Friedman, S.; Mor, C. The success of endodontic therapy—Healing and functionality. *J. Calif. Dent. Assoc.* **2004**, *32*, 493–503. [PubMed]
92. Regan, J.D.; Fleury, A.A. Irrigants in non-surgical endodontic treatment. *J. Ir. Dent. Assoc.* **2006**, *52*, 84–92.
93. Gilbert, P.; Moore, L.E. Cationic antiseptics: Diversity of action under a common epithet. *J. Appl. Microbiol.* **2005**, *99*, 703–715. [CrossRef] [PubMed]
94. Schwarz, S.R.; Hirsch, S.; Hiergeist, A.; Kirschneck, C.; Muehler, D.; Hiller, K.A.; Maisch, T.; Al-Ahmad, A.; Gessner, A.; Buchalla, W.; et al. Limited antimicrobial efficacy of oral care antiseptics in microcosm biofilms and phenotypic adaptation of bacteria upon repeated exposure. *Clin. Oral Investig.* **2021**, *25*, 2939–2950. [CrossRef]

95. Yesilsoy, C.; Whitaker, E.; Cleveland, D.; Phillips, E.; Trope, M. Antimicrobial and toxic effects of established and potential root canal irrigants. *J. Endod.* **1995**, *21*, 513–515. [CrossRef]
96. Kandaswamy, D.; Venkateshbabu, N.; Gogulnath, D.; Kindo, A.J. Dentinal tubule disinfection with 2% chlorhexidine gel, propolis, morinda citrifolia juice, 2% povidone iodine, and calcium hydroxide. *Int. Endod. J.* **2010**, *43*, 419–423. [CrossRef] [PubMed]

antibiotics

MDPI

Article

Attitudes of Dental Students towards the Prescription of Antibiotics during Endodontic Treatment

Lauzan haj Khalaf, Salma Kabbaj and Babacar Toure *

Department of Conservative Dentistry and Endodontics, Faculty of Dental Medicine, College of Health Sciences, International University of Rabat, Rabat 11103, Morocco; lauzan.haj.khalaf@uir.ac.ma (L.h.K.); salma.kabbaj@uir.ac.ma (S.K.)
* Correspondence: babacar.toure@uir.ac.ma

Abstract: Aim: This study aimed to evaluate the antibiotic-prescribing attitudes of dental students during the management of endodontic infections. Materials and methods: This study was conducted in the five faculties of dental medicine in Morocco. A self-administered questionnaire was used and completed online. This questionnaire has three parts: the first includes the socio-demographic data, the second is related to the types of antibiotics prescribed, and the final addresses clinical scenarios in which antibiotics are prescribed. Data were analyzed with Jamovi, and X_2 and ANOVA tests were performed. Results: Three hundred and sixty-five students participated in this study. The average duration of antibiotic therapy was 5.87 ± 1.45. Of all the students, 83.8% prescribe amoxicillin first. For patients with penicillin allergy, clindamycin was the most prescribed, amounting to 53.9%. Antibiotics are prescribed for all pulp and periapical pathologies. For acute pulpitis and acute apical periodontitis, a statistically significant difference between the different faculties was noted ($p = 0.03$). Regarding apical abscesses, antibiotic prescription was more frequent at the public faculty of Casablanca, corresponding to 92.8%. ($p = 0.02$). Conclusion: It appears from this study that there is a need for faculties to develop innovative teaching models to improve students' level of knowledge on antibiotics and their indications in endodontics.

Keywords: antibiotics; apical periodontitis; prescription; pulp pathologies; students

1. Introduction

According to the European Society of Endodontics, the use of antibiotics is recommended when an infection is persistent or systemic. Thus, for rational use, consensus conferences were organized to codify their indications [1]. Antibiotics are used in the following pathological situations: acute apical abscesses in medically compromised patients; patients at risk; and acute apical abscess with systemic involvement, i.e., localized swellings, fever > 38°, asthenia, insomnia, malaise, lymphadenopathy, trismus, progressive infections, rapid onset of severe infections, and dental reimplantation after expulsion [2]. Consequently, the prescription of antibiotics is not systematic in endodontics. Therefore, their inappropriate use can contribute to the occurrence of antibiotic resistance. The Lancet Infectious Diseases Commission has published a series of articles that sound the alarm on antibiotic resistance and encourage the community of practitioners to be more cautious when prescribing [3,4]. In this context, several medical professionals have started studies on the use of antibiotics. In endodontics, these studies have focused on general practitioners in several countries [5–8]. The results of these studies showed that very few practitioners use antibiotics appropriately in the management of pulpal and periapical pathologies [9–12]. An analysis of the literature shows that very few studies have focused on the attitudes and practices of students in clinical training. On this topic, according to the drug regimen, Mohanty et al. [13] reported that 30.85% of postgraduate students of endodontics prescribed pretreatment medication; in comparison, 62.40% of the analyzed endodontists preferred medication for both. European training program guidelines state that graduates should be adequately trained in the basic and clinical science of endodontics. They specify that students must not only have knowledge of the microbiology of pathologies of endodontic origin but also be competent enough to manage these pathologies. This management integrates infection control, pharmacology, and endodontic therapy.

In Morocco, the book of educational standards for dental studies mentions that dental students should know how to use antibiotics in managing microbial infections, the mechanisms of action, and the issues related to antibiotic resistance. They must also be able to perform simple root canal treatments. In clinical programs, students learn to apply their knowledge of antibiotics to manage infectious dental diseases. At the end of these courses, they must know the indications and contraindications for the use of antibiotics in endodontics.

Despite these guidelines in training programs, studies carried out among students in training in Spain reveal a lack of knowledge and inappropriate treatment regimens. They also show a serious need to improve knowledge in regard to the prescription of antibiotics [14].

In Morocco, no study has analyzed the practices of students concerning the use of antibiotics during endodontic treatments.

The objective of the present study was to assess the level of knowledge and prescription attitudes of students in the management of pulpal and periapical pathologies.

2. Materials and Methods

This study is based on a cross-sectional descriptive survey conducted from November 2021 to February 2022.

The study was approved by the Ethics Committee of the University Clinic of Dental Medicine of the International Faculty of the International University of Rabat (CUMD/FIMD 03/22). This study was conducted on all dental students from the 4th to the 6th year enrolled at 5 faculties of dental medicine in Morocco: The International University of Rabat (UIR), the Abulcassis University of Health Sciences (UIASS), the Mohammed 6th University of Health Science (UM6SS), the public faculty of Dental Medicine of Rabat (FMPR), and the public faculty of Dental Medicine Casablanca (FMPC).

The sole prerequisite for participation was to be a dental student in the 4th to 6th year of undergraduate studies. The questions were based on those asked in previous surveys developed in several counties [5–8].

A self-administered questionnaire (Table 1) based on models from previous studies was used [6–12]. This questionnaire has three parts.

Table 1. Antibiotic use in endodontic infections questionnaire administered to dental students' undergraduates in Morocco.

Gender	**Male** ☐	**Female** ☐			
Undergraduate Year	4th☐	5th☐		6th☐	
Faculty of Dental Medicine: FMPR ☐		UIR ☐	UIASS ☐	UM6SS ☐	FMPC ☐
(1) When systemic antibiotics are indicated, which antibiotic would you choose for the treatment of an endodontic infection in an adult, healthy patient with no medical allergies? (choose one answer):					
Amoxicillin	500 mg ☐	1 g ☐			
Amoxicillin + Clavulanic Acid	1 g\125 mg	_☐			
Azithromycin	250 mg ☐				
Clarithromycin	500 mg ☐				
Clindamycin	300 mg ☐				
Erythromycin	500 mg ☐				
Metronidazol + Spiramycin	_☐				
Other ______					
(2) For how many days would you prescribe an antibiotic treatment? ______					
(3) When systemic antibiotics are indicated, which antibiotic would you choose for the treatment of an endodontic infection in an adult, healthy patient with allergy to penicillin? (choose one answer):					

Table 1. *Cont.*

Gender	Male ☐	Female ☐
Azithromycin	250 mg ☐ _ 500 mg	
Clarithromycin	500 mg ☐ __ ☐	
Clindamycin	300 mg	
Erythromycin	_ ☐	
Metronidazole + spiramycin	_ ☐	
Other		
(4) In which of the following situations antibiotics are indicated?		
Reversible pulpitis ☐ Irreversible acute pulpitis ☐ Apical acute periodontitis ☐ Chronic apical periodontitis without sinus tract ☐ Chronic apical periodontitis with sinus tract ☐ Acute apical abscess ☐ Pulp necrosis ☐ Endodontic retreatment ☐ Endodontic surgery ☐ Post-operative pain ☐		

The first part includes the socio-demographic data of the students: gender, level of study, and dental medicine faculty of origin. The second part is related to the types of antibiotics prescribed during endodontic treatment of an adult patient with or without an allergy to penicillin and the duration of the prescription of these antibiotics.

The third and final part addresses clinical scenarios in which antibiotics are routinely prescribed. The students who took part did so on a voluntary basis, without any form of compensation, and in an anonymous manner.

This questionnaire was sent online via email; a reminder was sent every 15 days. Questionnaires received after February 2024 were not included in the statistical analysis.

Data were collected and analyzed with Jamovi version 1.8.1 (software). Chi-square test and ANOVA tests were performed to compare qualitative and quantitative variables. The significance level was set at $p < 0.05$.

3. Results

In total, 365 students responded to the survey; 71.9% were female, and 28.1% were male, which yields a sex ratio of 2.49 in favor of women. Sixth-year students were more represented, with a percentage of 48% (169 students), than fifth-year students, constituting 28.7% of the sample (101 students), and, finally, the cohort comprised 82 students, representing 23.3% of the total number of students in their fourth year. The International University of Rabat (UIR) was the faculty that participated the most in this study, with a percentage of 39.9% (144 students); then, in second place was the Abulcassis University of Health Sciences (UIASS), with a percentage of 19.9% (72 students), and the Mohammed 6th University of Health Science (UM6SS), with a percentage more or less equivalent to that of Abulcassis, i.e., 19.7% (71 students). The public faculty of Dental Medicine of Rabat (FMPR) accounted for a percentage of 11.1% (40 students), and Casablanca (FMPC) had a percentage of 7.8% (28 students) (Figure 1).

The average duration was 5.87 ± 1.45 days, with a maximum of 7 days for 56.5% of the students and a minimum of 3 days for 14% of the students. No statistical difference was found between the different universities ($p > 0.05$).

3.1. For a Patient without a Penicillin Allergy

Most of the students, 83.8%, prescribe amoxicillin as their first intention, specifically the 1 g form of amoxicillin, which is the most prescribed. Amoxicillin associated with clavulanic acid was prescribed by 9.2%, azithromycin was proposed by 2.5% of students, and clindamycin was chosen by 1.4%. Concerning the prescription of Metronidazole associated with Spiramycin, we note a prescription percentage of 3.1%.

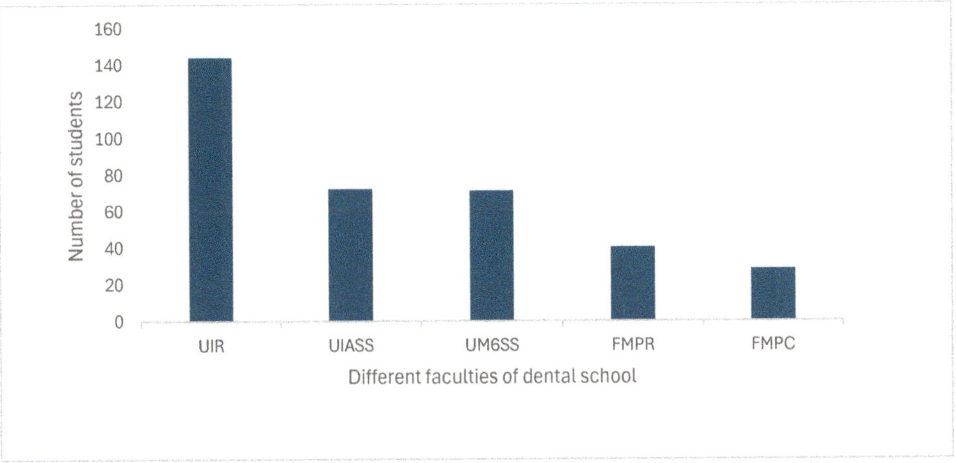

Figure 1. Distribution of the students according to the faculties of dental medicine. UIR, International University of Rabat; UIASS, Abulcassis University of Health Sciences; UM6SS, Mohammed 6th University of Health Sciences; FMPR, Public faculty of Dental Medicine of Rabat; FMPC, Public faculty of Dental Medicine of Casablanca.

3.2. For a Patient with a Penicillin Allergy

Concerning patients with a penicillin allergy, clindamycin was the most prescribed molecule (53.9%), followed by azithromycin (20.9%). Erythromycin was indicated in 7.8% of cases, and metronidazole combined with spiramycin was indicated in 17.4%.

3.3. Endodontic Pathologies with Antibiotic Prescription

Antibiotics are prescribed for all pulp and periapical pathologies, whether acute or chronic (Table 2). For acute reversible and irreversible pulpitis, antibiotic prescriptions were more indicated by fourth-year students than by students in higher years. In cases of acute reversible pulpitis (21), 2% of fourth-year students prescribed antibiotics, compared to 9.9% of fifth-year students and 2.9% in their sixth year. This difference was statistically significant ($p = 0.01$).

Table 2. Distribution of students who prescribe antibiotics according to endodontic pathologies.

N = Number of Prescriptions	**4th**	**5th**	**6th**	
% = percentage	N (%)	N (%)	N (%)	*p*
Reversible acute pulpitis	17 (21.2)	10 (9.9)	5 (2.9)	0.01 *
Irreversible acute pulpitis	14 (17.5)	8 (7.9)	11 (6.5)	0.06
Acute apical periodontitis	11 (13.7)	20 (19.8)	31 (18.4)	0.49
Chronic apical periodontitis	13 (16.2)	15 (14.8)	25 (14.8)	0.9
Apical periodontitis with fistula	32 (40)	43 (42.5)	80 (47.6)	0.67
Acute apical abscess	53 (66.2)	74 (73.2)	134 (79.7)	0.13
Pulp necrosis	11(13.7)	6 (5.9)	5 (2.9)	0.005 *
Endodontic retreatment	11 (13.7)	4 (3.6)	10 (5.9)	0.03 *
Endodontic surgery	31 (38.7)	48 (47.5)	88 (52.3)	0.11
Post op pain	16 (20)	16 (15.8)	25 (14.8)	0.47

* significant; N = number of prescriptions; % = percentage.

For periapical pathologies, antibiotics were more indicated for acute apical abscesses, with percentages reaching 76.6%. No statistically significant difference was noted between the study levels. Regarding pain following endodontic treatment, antibiotic prescriptions were recommended by 20% of fourth-year students, 15.8% of fifth-year students, and

14.8% of sixth-year students; this difference was not statistically significant. However, a statistically significant difference was observed regarding the prescription of antibiotics in cases of pulp necrosis and endodontic retreatment. Students in their fourth year were found to have a higher rate of prescription than their counterparts at other levels.

The analysis of the results according to the different universities shows statistically significant differences concerning the prescription of antibiotics for acute apical abscesses, acute apical periodontitis, and endodontic surgery (Table 3).

Table 3. Distribution of students who prescribe antibiotics for endodontic pathologies according to university.

	UIR N = 142	**UIASS** N = 72	**UM6SS** N = 70	**FMPC** N = 28	**FMPR** N = 40	*p*
	N (%)	N (%)	N (%)	N (%)	N (%)	
Reversible pulpitis	20 (14)	2 (2.7)	7 (10)	0	3 (7.5)	0.32
Acute pulpitis irreversible	11 (7.7)	7 (9.7)	12 (17.1)	0	1 (2.5)	0.47
Acute apical periodontitis	14 (9.8)	12 (16.6)	13 (18.5)	8 (28.5)	7 (17.5)	0.03 *
Apical periodontitis with fistula	58 (40.8)	30 (41.6)	38 (54.2)	18 (64.2)	16 (40)	0.11
Acute apical abscess	95 (66.9)	56 (77.7)	55 (78.5)	26 (92.8)	33 (82.5)	0.02 *
Phoenix abscess	53 (37.3)	28 (38.8)	30 (42.8)	18 (64.2)	19 (47.5)	0.03 *
Pulp necrosis	11 (7.7)	2 (2.7)	5 (7.1)	2 (7.1)	1 (2.5)	0.43
Endodontic retreatment	8 (5.6)	7 (9.7)	6 (8.5)	2 (7.1)	2 (5%)	0.8
Endodontic surgery:	52 (36.6)	41 (56.9)	44 (62.8)	16 (51.1)	15 (37.5)	0.004 *
Post op pain	20 (14)	13 (18)	16 (22.8)	6 (21.4)	3 (7.5)	0.6

* significant; N = number of prescriptions; % = percentage. UIR, International University of Rabat; UIASS, Abulcassis University of Health Sciences; UM6SS, Mohammed 6th University of Health; FMPR, Public faculty of Dental Medicine of Rabat; FMPC, Public faculty of Dental Medicine of Casablanca.

The percentage of students who prescribed antibiotics in cases of apical periodontitis was significantly high at the Public Faculty of Casablanca (FMPC), amounting to 28.5%, compared to the percentages at other faculties (UM6SS, 18.5%; FMPR, 17.5%; and UIR, 9.8%, with $p = 0.03$).

For acute apical abscesses, the prescription of antibiotics was also higher among the students of the Public Faculty of Casablanca (92.8%) than at the other faculties ($p = 0.02$).

4. Discussion

This study deals with the knowledge and practices of students concerning the use of antibiotics during the management of pulpal and periapical pathologies. The questionnaire used in this study was proposed in previous studies carried out in several counties [12,14,15].

The sample consisted of 365 students from five faculties of dentistry in Morocco. Three of these faculties are in Rabat, namely, the International University of Rabat (UIR), Abulcassis University of Health Science (UIASS), and the Public University of Rabat (FMPR), and two of these faculties are in Casablanca, i.e., the Mohammed 6 University of Health Science (UM6SS) and the Public Faculty of Casablanca (FMPC).

Our sample is fairly representative because in the average percentage of participation recorded in similar studies [15,16], the number of women was higher (71.9%). This difference was also noted in Spain, which shows a feminization of the profession. The noted prescription period of 5.87 ± 1.45 days is in line with the recommendations because endodontic infections always regress within 3 to 7 days if the infectious cause is eliminated [17]. For patients without penicillin allergy, amoxicillin is the molecule of choice for students, with 83.8% selecting this antibiotic, followed by the combination of amoxicillin and clavulanic acid, at 9.2%. These results corroborate the data found in the study by Khaloufi et al., carried out among practitioners in Northern Morocco [15]. Furthermore, these results are comparable to those observed in other European countries [16] and Africa [18].

In India, practitioners also prefer amoxicillin as their first choice, followed by oxofloxacin. Amoxicillin is an effective antibiotic for the germs implicated in periapical pathologies.

However, in the event of inefficiency linked to the production of β-lactamase, the combination of amoxicillin with clavulanic acid can be proposed. According to scientific societies, this combination should be considered a second-line treatment in the event of failure using amoxicillin or for patients with established immunity [1]. For patients with a penicillin allergy, clindamycin was the most prescribed molecule (53.9%), followed by azithromycin (20.9%). Erythromycin was indicated in 7.8% of cases, and metronidazole combined with spiramycin was indicated in 17.4%. The study by Bolfini et al. showed results identical to this study, where clindamycin was the most prescribed antibiotic, at 33%. Rodriguez Nunez et al. [19] also reported clindamycin to be in first place, at 69%, followed by azithromycin, at 29.2%. In contrast, Al Khuzaei et al. [7], in a study carried out on dental surgeons, found different results, with azithromycin being at the top of the line, with 63.2%. Metronidazole occupies third place in this study. It is an anti-infective effective against bacteria with black pigmentation, but it is less effective on aerobes or facultative anaerobes, which makes it more advisable for it to be combined with another antibiotic such as amoxicillin or spiramycin. Its combination with amoxicillin should not be systematic but dictated by the evolution of the pathology; if there is no favorable evolution two to three days after the prescription of amoxicillin alone, metronidazole can be added to amoxicillin. In Spain, 99% of students chose clindamycin. The same study carried out among endodontists and general practitioners found different results (63% and 65% for clindamycin, respectively).

Regarding the indications of antibiotics as an adjuvant to endodontic treatment, several studies carried out in different countries show a lack of knowledge on indications and the inappropriate use of antibiotics. In this study, the percentage of students who prescribe antibiotics is relatively high. The results show that 9.1% of students prescribe antibiotics for acute pulpitis. A similar study carried out among students in Spain showed higher results for pulpitis (29%) and irreversible pulpitis with symptomatic apical periodontitis and moderate/severe symptoms (63%) [14]. However, these pathologies are dominated by pulpal inflammation; no trace of infection is noted in the pulp. Their management, therefore, does not require the use of antibiotics.

For chronic apical periodontitis, 15.4% of students prescribe antibiotics. These results are similar to those found in Spain (16%) [14]. This decrease in percentage could be mainly related to the chronic asymptomatic nature of these cases. However, compared to the results obtained by practitioners, the percentages are higher, with 31% in Spain and 51.2% in Morocco [15].

These results show the need for the continuous training of general practitioners.

In the case of retreatment, Moroccan students, like most dentists in India, prefer to prescribe antibiotics only in specific cases [20,21]. Otherwise, the use of a solvent during root canal retreatment does not cause any significant difference in the post-operative pain levels or medication intake for the retrieval of Gutta-percha [22].

Concerning postoperative pain, antibiotics were prescribed by dental students with a rate ranging from 7.5% to 22.8% (according to dental school). Therefore, Jose et al. [23] concluded that oral consumption of corticosteroids is a better analgesic in this case.

For acute apical abscesses, 76.6% of students prescribe antibiotics. In Spain [5,19], it is almost systematic, at 90%. Indeed, the prescription of an antibiotic should only be indicated in the presence of associated general signs [1,24].

5. Conclusions

The results of the present study show that despite the existence of pharmacology and endodontics modules in study curricula, the various faculties of dentistry should integrate new teaching methods to improve students' knowledge of antibiotics and their indications in endodontics. They must develop innovative interactive teaching approaches based on real cases, electronic educational tools offering access to precise information, and standardized educational materials for prudent antibiotic therapy. The curriculum for dental studies should place more emphasis on prescription and the teaching of good practices.

Author Contributions: Conceptualization, B.T.; methodology, L.h.K. and B.T.; investigation, S.K. All authors have read and agreed to the published version of the manuscript.

Funding: This research received no external funding.

Institutional Review Board Statement: This study was conducted in accordance with the Declaration of Helsinki and approved by The Research Ethics Board of the International Faculty of Dental Medicine of Rabat—International University of Rabat Ref: CUMD/FIMD 003/20/22/Approval/2019.

Informed Consent Statement: Informed consent was obtained from each author.

Data Availability Statement: Data is contained within the article.

Conflicts of Interest: The authors declare no conflict of interest.

References

1. Segura-Egea, J.J.; Gould, K.; Hakan Şen, B.; Jonasson, P.; Cotti, E.; Mazzoni, A.; Sunay, H.; Tjäderhane, L.; Dummer, P.M.H. European Society of Endodontology position statement: The use of antibiotics in endodontics. *Int. Endod. J.* **2018**, *511*, 20–25. [CrossRef]
2. Hinckfuss, S.E.; Messer, L.B. An evidence-based assessment of the clinical guidelines for replanted avulsed teeth. Part 2: Prescription of systemic antibiotics. *J. Dental Traumatol.* **2009**, *25*, 158–164. [CrossRef] [PubMed]
3. Jenks, J. Antibiotic resistance needs global solutions. *Lancet Infect. Dis.* **2014**, *147*, 550. [CrossRef] [PubMed]
4. Howard, S.J.; Catchpole, M.; Watson, J.; Davies, S.C. Antibiotic resistance: Global response needed. *Lancet Infect Dis.* **2013**, *13*, 1001–1003. [CrossRef]
5. Segura-Egea, J.J.; Velasco-Ortega, E.; Torres-Lagares, D.; Velasco-Ponferrada, M.C.; Monsalve-Guil, L.; Llamas-Carreras, J.M. Pattern of antibiotic prescription in the management of endodontic infections among Spanish oral surgeons. *Int. Endod. J.* **2010**, *43*, 342–350. [CrossRef] [PubMed]
6. Bolfoni, M.R.; Pappen, F.G.; Pereira-cenci, T.; Jacinto, R.C. Antibiotic prescription for endodontic infections: A survey of Brazilian Endodontists. *Int. Endod. J.* **2018**, *512*, 148–156. [CrossRef] [PubMed]
7. Al Khuzaei, N.M.; Assery, M.K.; Al Rahbeni, T. Knowledge of antibiotics among dentists in Saudi Arabia. *J. Int. Oral. Health* **2017**, *92*, 1198–1203.
8. Kakoei, S.; Raoof, M.; Baghaei, F.; Adhami, S. Pattern of antibiotic prescription among dentists in Iran. *Iran Endod. J.* **2007**, *2*, 19–23.
9. Iqbal, A. The Attitudes of Dentists towards the Prescription of Antibiotics during Endodontic Treatment in North of Saudi Arabia. *J. Clin. Diagn. Res.* **2015**, *9*, ZC82–ZC84. [CrossRef]
10. Alonso-Ezpeleta, O.; Milagros, M.J.; Martin-Biedma, B. Use of antibiotics by spanish dentists receiving postgraduate training in endodontics. *J. Clin. Exp. Dent.* **2018**, *107*, 687–695. [CrossRef]
11. Atif, M.; Azeem, M.; Rehan Sarwar, M.; Malik, I.; Ahmad, W.; Hassan, F.; Rehman, A.; Rana, M. Evaluation of prescription errors and prescribing indicators in the private practices in Bahawalpur, Pakistan. *J. Chin. Med. Assoc.* **2018**, *81*, 444–449. [CrossRef]
12. Yingling, N.M.; Byme, B.E.; Hartwell, G.R. Antibiotic use by members of the American Association of Endodontists in the year 2000: Report of a national survey. *J. Endod.* **2002**, *285*, 396–404. [CrossRef] [PubMed]
13. Mohanty, A.; Patro, S.; Barman, D.; Jnaneswar, A. Modern endodontic practices among dentists in India: A comparative cross-sectional nation-based survey. *J. Conserv. Dent.* **2020**, *23*, 441–446.
14. Martín-Jiménez, M.; Martín-Biedma, B.; López-López, J.; Alonso-Ezpeleta, O.; Velasco-Ortega, E.; Jiménez-Sánchez, M.C.; Segura-Egea, J.J. Dental students' knowledge regarding the indications for antibiotics in the management of endodontic infections. *Int. Endod. J.* **2018**, *511*, 118–127. [CrossRef]
15. Khalloufi, O.; Haj Khalaf, L.; Akerzoul, N.; Ibn Majdoub Hassani, F.Z.; Touré, B. Attitudes of dental practitioners from Northern Morocco on the prescription of antibiotics during endodontic treatment: A survey. *Saudi Endod. J.* **2022**, *12*, 316–321.
16. Licata, F.; Di Gennaro, G.; Cautela, V.; Nobile, C.G.A.; Bianco, A. Endodontic Infections and the Extent of Antibiotic Overprescription among Italian Dental Practitioners. *Antimicrob. Agents Chemother.* **2021**, *65*, e0091421. [CrossRef]
17. Pallasch, T.J. How to use antibiotics effectively. *J. Calif. Dent. Assoc.* **1993**, *21*, 46–50.
18. Lalloo, R.; Solanki, G.; Ramphoma, K.; Myburgh, N.G. Endodontic treatment-related antibiotic prescribing patterns of South African oral health professionals. *Int. Endod. J.* **2017**, *50*, 1027–1033. [CrossRef]
19. Rodriguez-Núñez, A.; Cisneros-Cabello, R.; Velasco-Ortega, E.; Llamas-Carreras, J.M.; Tórres-Lagares, D.; Segura-Egea, J.J. Antibiotic use by members of the Spanish Endodontic Society. *J. Endod.* **2009**, *35*, 1198–1203. [CrossRef] [PubMed]

20. Mandke, L.; Koparkar, T.; Bhagwat, S.; Vimala, N.; Vandekar, M. Endodontic retreatment practice trends among dental surgeons: Survey-based research. *J. Conserv. Dent. Endod.* **2023**, *26*, 663–670.
21. Pandey, V.; Dey, S. Antibiotic usage: A survey among prescribing habits of members of Indian endodontic society, davange Rebranch. *Guident* **2014**, *7*, 62–63.
22. Shah, T.; Ramesh, S.; Sugumaran, S.; Choudhari, S. Endodontic retreatment efficacy with and without solvents: A systematic review. *J. Conserv. Dent. Endod.* **2023**, *26*, 610–615. [PubMed]
23. Jose, J.; Teja, K.V.; Palanivelu, A.; Khandelwal, A.; Siddique, R. Analgesic efficacy of corticosteroids and nonsteroidal anti-inflammatory drugs through oral route in the reduction of postendodontic pain: A systematic review. *J. Conserv. Dent. Endod.* **2022**, *25*, 9–19. [CrossRef] [PubMed]
24. Solomonov, M.; Batashvili, G.; Shuster, A.; Slutzky, H.; Moshonov, J.; Buchkovskii, O.; Lvovsky, A.; Azizi, H.; Levin, A.; Ben Itzhak, J.; et al. Systemic antibiotics in endodontics-International questionnaire study: Choice of types, dosage, loading and duration. *Aust. Endod. J.* **2023**, *49*, 58–63. [CrossRef] [PubMed]

antibiotics

MDPI

Systematic Review

The Microbiome of Peri-Implantitis: A Systematic Review of Next-Generation Sequencing Studies

Koay Chun Giok [1] and Rohit Kunnath Menon [2,*]

[1] School of Dentistry, International Medical University, Kuala Lumpur 57000, Malaysia; koay.chungiok@student.imu.edu.my
[2] College of Dentistry, Ajman University, Ajman 346, United Arab Emirates
* Correspondence: r.menon@ajman.ac.ae

Abstract: (1) Introduction: Current evidence shows that mechanical debridement augmented with systemic and topical antibiotics may be beneficial for the treatment of peri-implantitis. The microbial profile of peri-implantitis plays a key role in identifying the most suitable antibiotics to be used for the treatment and prevention of peri-implantitis. This systematic review aimed to summarize and critically analyze the methodology and findings of studies which have utilized sequencing techniques to elucidate the microbial profiles of peri-implantitis. (2) Results: *Fusobacterium, Treponema, and Porphyromonas* sp. are associated with peri-implantitis. *Veillonella* sp. are associated with healthy implant sites and exhibit a reduced prevalence in deeper pockets and with greater severity of disease progression. *Streptococcus* sp. have been identified both in diseased and healthy sites. *Neisseria* sp. have been associated with healthy implants and negatively correlate with the probing depth. Methanogens and AAGPRs were also detected in peri-implantitis sites. (3) Methods: The study was registered with the International Prospective Register of Systematic Reviews (PROSPERO) (CRD42023459266). The PRISMA criteria were used to select articles retrieved from a systematic search of the Scopus, Cochrane, and Medline databases until 1 August 2023. Title and abstract screening was followed by a full-text review of the included articles. Thirty-two articles were included in the final qualitative analysis. (4) Conclusions: A distinct microbial profile could not be identified from studies employing sequencing techniques to identify the microbiome. Further studies are needed with more standardization to allow a comparison of findings. A universal clinical parameter for the diagnosis of peri-implantitis should be implemented in all future studies to minimize confounding factors. The subject pool should also be more diverse and larger to compensate for individual differences, and perhaps a distinct microbial profile can be seen with a larger sample size.

Keywords: peri-implantitis; microbiome; sequencing; dental implant; complications

1. Introduction

Dental implants exhibit high success rates of up to 97% and above [1]. However, contributory factors related to occlusal overloading and peri-implant tissue infection may lead to implant failure [2]. Peri-implantitis is defined as an infection of the peri-implant tissues accompanied by suppuration and clinically significant progressive crestal bone loss after the adaptive phase, leading to decreased osseointegration and pocket formation [3,4]. Peri-implantitis has a reported prevalence ranging from 6.6% to 51% [5–9]. Various risk factors are associated with an increased risk of peri-implantitis. Prosthetic factors, including convex emergence profiles, submucosal crown margins, and excess cement in cemented implant prostheses, increase the risk of peri-implantitis [2,3]. Systemic conditions such as diabetes mellitus and osteoporosis also increase the risk of peri-implantitis [10]. Furthermore, smoking has been found to directly affect the bone surrounding the implant, thereby increasing the risk of peri-implantitis as well [11]. Biofilm removal and control with instruments such as Gracey curettes, ultrasonic scalers, and air powder abrasive devices have been employed with questionable success in the treatment of peri-implantitis since mechanical debridement also comes with its challenges, especially at the apically facing thread surfaces, as demonstrated by Steiger-Ronay et al. [12]. Antimicrobials are also ineffective if mechanical debridement is inadequately performed, as mentioned previously [13,14]. However, liquid desiccants have been reported to reduce the anaerobic bacteria load in diseased implants [15]. To date, the treatment of peri-implantitis is similar to that of periodontitis [16]. The prognosis of this condition is uncertain, and hence, determining the fundamental cause is important for preventive strategies and also targeted approaches [17].

The exact mechanism of microbial interaction in peri-implantitis is not clearly known [3]. Initial studies reported that Staphylococcus aureus plays a role in the progression of the disease [18,19]. However, the consensus on the predominance of S. aureus in peri-implantitis sites was contradicted by Belibasakis et al., as their study concluded the predominance of Treponema spp. and Synergistetes cluster A in peri-implantitis sites [19,20].

Koyanagi et al. reported a more diverse microbial profile compared to that of periodontitis [21], while other studies indicated similarity [22,23]. A microbial profile consisting of aggressive and resistant microorganisms distinct from periodontitis has also been reported previously [24]. Periodontally involved teeth act as reservoir for periopathogens which translocate to the implant sites, making chronic periodontitis an important risk factor for peri-implantitis [21,23,25,26].

Culture-dependent studies evaluating the microbiome of peri-implantitis have limited insights into the bacterial community [27,28], and more recent next-generation sequencing techniques may give us an insight into a more targeted approach to peri-implantitis treatment which, in turn, can improve the prognosis of this condition [29]. The use of next-generation sequencing allows the identification of non-culturable species as compared to conventional methods [29]. The detection of bacterial and fungal infections has been shown to be consistently accurate as compared to conventional methods [30]. In addition, next-generation sequencing has been shown to be cost-effective for identifying the disease with a given high pretest probability, as compared to culture methods [31].

This systematic review aims to summarize and critically analyze the methodology and findings of studies that have utilized next-generation sequencing techniques to elucidate the microbial profiles of peri-implantitis.

2. Results

From the initial search, 506 articles were identified after the elimination of duplicates. After performing the preliminary review of the title and abstracts, 32 articles were included for full-text screening. Based on the selection criteria, 32 studies were chosen to be included in the qualitative analysis (Figure 1). The Risk Of Bias In Non-randomized Studies–of Exposures (ROBINS-E) assessment of 32 articles is shown in Table 1. The Grading of Recommendations, Assessment, Development, and Evaluation (GRADE) approach was used (Table 2) and revealed a low certainty of evidence for the outcomes of diversity and richness as well as the abundance of taxa.

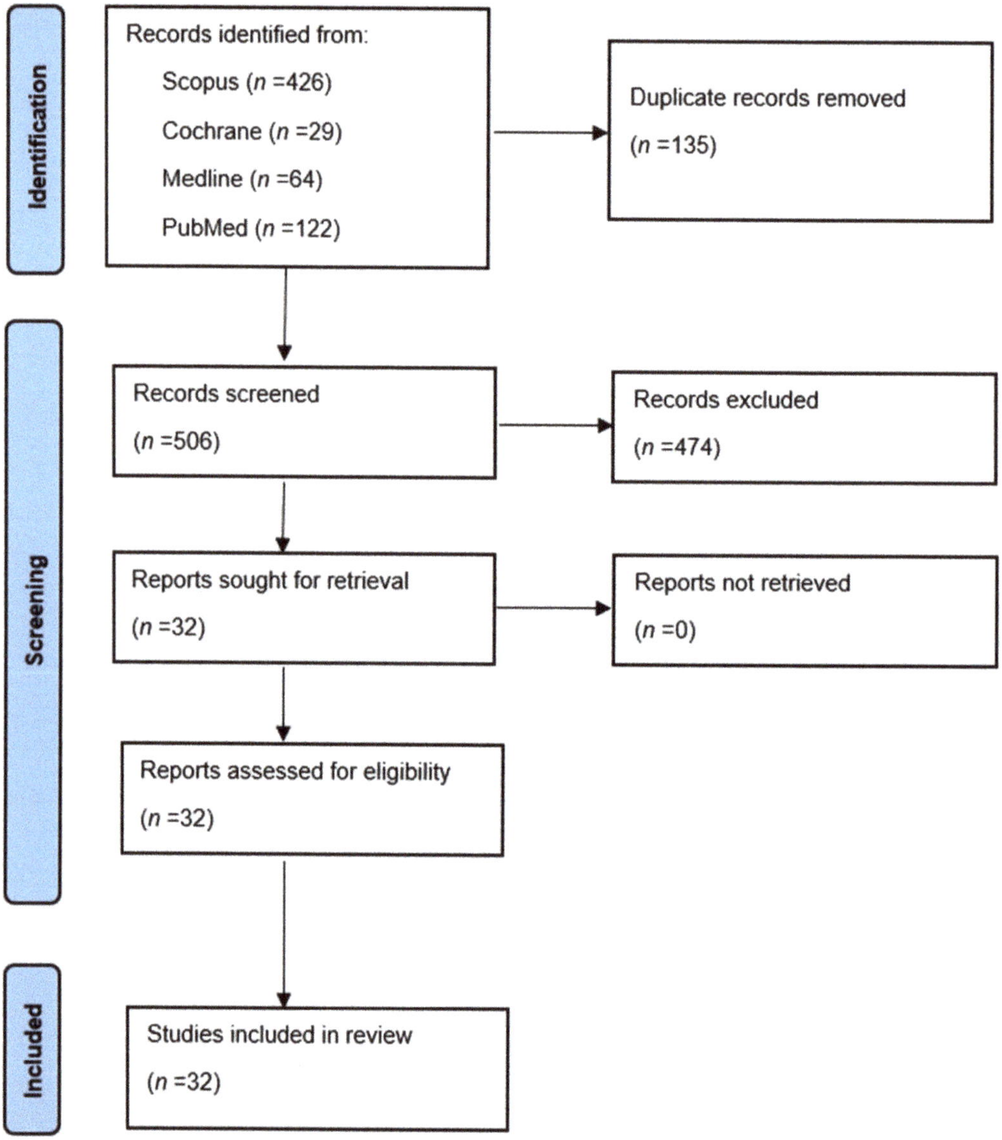

Figure 1. PRISMA flowchart.

Table 1. The Risk Of Bias In Non-randomized Studies–of Exposures (ROBINS-E) assessment.

Author, Year	Confounding Variables	Measurement of the Exposure	Selection of Participants	Post-Exposure Interventions	Missing Data	Measurement of Outcome	Selection of Reported Result	Overall Bias
Kim et al., 2023 [32]	S	L	S	S	L	L	L	S
Song et al., 2022 [33]	L	L	S	L	L	L	L	S
Pallos et al., 2022 [34]	H	L	L	L	L	L	L	H
Barbagallo et al., 2022 [35]	H	S	S	L	L	L	L	H
Shi et al., 2021 [36]	S	L	L	L	L	L	L	L
Polymeri et al., 2021 [37]	L	L	L	L	L	L	L	L
Korsch et al., 2021 [38]	L	L	L	L	L	L	L	L
Komatsu et al., 2020 [39]	S	L	L	L	L	L	L	L
Ghensi et al., 2020 [40]	S	S	S	L	L	L	L	S
Aleksandrowicz et al., 2020 [41]	S	L	L	L	L	L	L	L
Yu et al., 2019 [42]	S	L	L	L	L	L	L	L
Kröger et al., 2018 [43]	L	L	H	L	L	L	S	H
Gao et al., 2018 [44]	L	L	L	L	L	L	L	L
Daubert et al., 2018 [45]	L	S	L	L	L	L	L	S
Al-Ahmad et al., 2018 [46]	L	L	L	L	L	L	L	L
Sousa et al., 2016 [47]	L	L	L	L	L	L	L	L
Sanz-Martin et al., 2017 [20]	L	L	L	L	L	L	L	L
Apatzidou et al., 2017 [23]	S	S	L	L	L	L	L	S
Yu et al., 2016 [48]	S	L	L	L	L	L	L	L
Shiba et al., 2016 [49]	S	L	S	L	L	L	L	S
Tsigarida et al., 2015 [50]	L	L	L	L	L	L	L	L
Jakobi et al., 2015 [51]	S	L	S	L	L	L	L	S
Zheng et al., 2014 [52]	L	L	L	L	L	L	L	L
Schaumann et al., 2014 [53]	S	L	S	L	L	L	L	S
Maruyama et al., 2014 [54]	S	S	S	L	L	L	L	S

Table 1. *Cont.*

Author, Year	Confounding Variables	Measurement of the Exposure	Selection of Participants	Post-Exposure Interventions	Missing Data	Measurement of Outcome	Selection of Reported Result	Overall Bias
Tamura et al., 2013 [55]	L	L	L	L	L	L	L	L
Koyanagi et al., 2013 [21]	S	L	L	L	L	L	L	L
Dabdoub et al., 2013 [25]	L	L	L	L	L	L	L	L
da Silva et al., 2013 [56]	L	L	L	L	L	L	L	L
Kumar et al., 2012 [22]	H	S	S	L	L	L	L	H
Koyanagi et al., 2010 [57]	S	L	L	L	L	L	L	L
Faveri et al., 2010 [58]	L	L	L	L	L	L	L	L

L: low risk of bias; S: some concerns; H: high risk of bias.

Table 2. Grading of Recommendations, Assessment, Development, and Evaluation (GRADE) approach.

<table>
<tr><th colspan="7">Certainty Assessment</th><th colspan="3">Summary of Findings</th></tr>
<tr><th rowspan="2">Participants (Studies) Follow-Up</th><th rowspan="2">Risk of Bias</th><th rowspan="2">Inconsistency</th><th rowspan="2">Indirectness</th><th rowspan="2">Imprecision</th><th rowspan="2">Publication Bias</th><th rowspan="2">Overall Certainty of Evidence</th><th colspan="2">Study Event Rates (%)</th><th rowspan="2">Impact</th></tr>
<tr><th>With Conventional Methods</th><th>With Next-Generation Sequencing</th></tr>
<tr><td colspan="10">Outcome: Diversity and Richness</td></tr>
<tr><td>1069 (32 observational studies)</td><td>serious [a]</td><td>serious [b]</td><td>serious [c]</td><td>not serious</td><td>All plausible residual confounding would reduce the demonstrated effect</td><td>Low</td><td colspan="3">The diversity and richness of the microbiome is heterogeneous and inconsistent across all 32 studies.</td></tr>
<tr><td colspan="10">Outcome: Abundance of Taxa</td></tr>
<tr><td>1069 (32 observational studies)</td><td>serious [a]</td><td>serious [b]</td><td>serious [c]</td><td>not serious</td><td>All plausible residual confounding would suggest a spurious effect, while no effect was observed</td><td>Low</td><td colspan="3">A heterogeneous pattern of taxa can be seen across all 32 studies reviewed.. The evidence suggests that next-generation sequencing has detected previously uncultured bacteria in diseased sites.</td></tr>
</table>

[a]. Out of the 32 studies reviewed, nine were of some concern, while four were at a high risk of bias based on the ROBINS-E assessment tool. [b]. Inconsistency is seen due to the heterogeneity across all 32 studies. [c]. Indirectness is seen due to the differences in the severity of peri-implantitis. Microbial compositions of different severities present heterogenous results.

2.1. Methodology of Studies

The methodological characteristics of the studies published between 2009 and 2021 are depicted in Table 3. The total sample size of the selected studies ranged from two to one hundred and six. Fifteen studies compared the association of the periodontitis microbiome with the peri-implantitis site microbiome [21,22,25,32,35,39,41,42,45,47–49,51,54,57]. Twelve studies compared the microbiomes of healthy implant (HI) sites to those of peri-implantitis (PI) sites [20,33,34,40,43,44,46,51,53,55,56,58], where the healthy implant site was the control. Peri-implant mucositis (PM) was also compared to peri-implantitis in seven studies [36,37,40,47,50,52]. Smoking was investigated as a factor in microbial dysbiosis in two studies [49,50]. Furthermore, Kroger et al. [43] investigated the association between the microbial diversity and the pocket depths of implants, while Korsh et al. [38] investigated the microbiota associated with early versus late implant loss.

Oral samples collected for microbiome isolation in the 32 included studies were composed mostly of subgingival plaque samples [20–23,25,32,33,35–58]. Two studies utilized supragingival plaque samples [32,53]. Sterile paper points were used to collect the subgingival plaque samples [21–23,25,33,35–39,42–44,46,48–51,53–55,57]. Eight studies utilized sterile Gracey curettes [20,32,40,41,45,47,56,58], while one study used a periodontal probe [52]. Further details on the collection method are provided in Table 3.

The DNA extraction technique, sequencing technique, targeted region, and the reference database for each study are summarized in Table 4. The microbiome profile is depicted in relation to the diversity, richness, and taxa abundance in Table 5.

Among the 32 studies reviewed, seven studies found an increase in the microbial diversity of peri-implantitis sites as compared with healthy implant sites [20,23,33,38,43,44,52]. Five studies did not report the diversity and richness of the samples collected [41,46,51,55,56,58]. Five studies reported an increase in the microbial diversity in peri-implantitis sites as compared with periodontitis sites [21,32,35,39,57]. Five studies reported a reduced microbial diversity in peri-implantitis sites compared with healthy implants in subgingival plaque [22,34,44,45,52]. Additionally, four studies reported no significant difference in diversity between healthy implants and peri-implantitis samples [23,33,37,50].

Table 3. Characteristics of the population and the results derived from the included studies.

Author, Year	Number of Subjects	Number of Implants	Study Setting	Duration of Implant	Case Definition for Peri-Implantitis/Peri-Implant Mucositis	Samples Collected	Collection Method
Kim et al., 2023 [32]	109	30 H, 30 PI	Korea	Not stated	PD $\geq$ 6 mm BOP Radiographic bone loss $\geq$3 mm	Supra- and subgingival plaque	Sterile Gracey curette
Song et al., 2022 [33]	14	14 H, 14 PI	China	Not stated	PD $\geq$ 6 mm Radiographic bone loss $\geq$3 mm	Subgingival plaque	Sterile paper point
Pallos et al., 2022 [34]	42	21 H, 21 PI	Brazil	$\geq$2 years	PD $\geq$ 5 mm BOP $\pm$ suppuration Radiographic bone loss $\geq$3 mm	Unstimulated saliva	Sterile plastic tube
Barbagallo et al., 2022 [35]	24	10 H, 24 PI	Italy	$\geq$1 year	Increasing PD since loading Evidence of radiographic bone loss BOP	Subgingival plaque	Sterile paper point
Shi et al., 2021 [36]	64	27 PM, 37 PI	China	$\geq$1 year	PD $\geq$ 6 mm BOP/suppuration Marginal bone loss $\geq$3 mm	Subgingival plaque	Sterile paper point
Polymeri et al., 2021 [37]	41	41 PI	The Netherlands	$\geq$1 year	PD $\geq$ 6 mm Clinical inflammation Radiographic bone loss $\geq$3 mm	Subgingival plaque	Sterile paper point
Korsch et al., 2021 [38]	48	31 PI, 22 H	Germany	$\leq$3 months or $\geq$3 years	PD $\geq$ 6 mm BOP and suppuration Radiographic bone loss $\geq$6 mm	Subgingival plaque	Sterile paper point
Komatsu et al., 2020 [39]	21	21 PI	Japan	$\geq$1 year	PD $\geq$ 6 mm BOP $\pm$ suppuration Radiographic bone loss $\geq$3 mm	Subgingival plaque	Sterile paper point
Ghensi et al., 2020 [40]	72	35 H, 37 PM, 41 PI	Italy	$\geq$1 year	BOP Radiographic bone loss > 2 mm	Subgingival plaque	Sterile Gracey curette

Table 3. *Cont.*

Author, Year	Number of Subjects	Number of Implants	Study Setting	Duration of Implant	Case Definition for Peri-Implantitis/Peri-Implant Mucositis	Samples Collected	Collection Method
Aleksandrowicz et al., 2020 [41]	139	37 H, 41 PI	Poland	Not stated	PD > 4 mm BOP Suppuration Visible three-thread loss	Subgingival plaque	Sterile Gracey curette
Yu et al., 2019 [42]	18	18 PI, 18 H	China	Not stated	PD $\geq$ 5 mm BOP and radiographic bone loss	Subgingival/submucosal plaque	Sterile paper point
Kröger et al., 2018 [43]	30	45 PI	Germany	Not stated	PD $\geq$ 5 mm BOP Radiographic bone loss $\geq$3 mm	Subgingival plaque	Sterile paper point
Gao et al., 2018 [44]	40	20 H, 20 PI	China	$\geq$6 months	PD $\geq$ 4 mm BOP Radiographic bone loss $\geq$2 mm	Subgingival plaque	Sterile paper point
Daubert et al., 2018 [45]	9	5 H, 6 PI	USA	Not stated	PD $\geq$ 4 mm BOP $\pm$ suppuration Radiographic bone loss > 2 mm	Subgingival plaque	Sterile 1/2 mini Gracey curette
Al-Ahmad et al., 2018 [46]	10	10 H, 10 PI	Germany	Not stated	PD $\geq$ 5 mm BOP and radiographic bone loss	Subgingival plaque	Sterile paper point
Sousa et al., 2016 [47]	18	2 H, 2 PM, 2 PI	UK	Not stated	PD $\geq$ 5 mm Radiographic bone loss of more than three threads up to half of the implant length or $\geq$2.5 mm BOP	Subgingival plaque	Sterile Gracey curette
Sanz-Martin et al., 2017 [20]	67	35 PI, 32 H	Switzerland	$\geq$1 year	Radiographic bone loss $\geq$2 mm at the mesial/distal side BOP	Subgingival plaque	Sterile Gracey curette
Apatzidou et al., 2017 [23]	10	4 H, 10 PI	Greece	$\geq$1 year	PD $\geq$ 6 mm BOP/suppuration Radiographic bone loss $\geq$2 mm	Subgingival plaque	Sterile paper point

Table 3. *Cont.*

Author, Year	Number of Subjects	Number of Implants	Study Setting	Duration of Implant	Case Definition for Peri-Implantitis/Peri-Implant Mucositis	Samples Collected	Collection Method
Yu et al., 2016 [48]	18	18 PI, 18 H	China	Not stated	PD $\geq$ 5 mm BOP and radiographic bone loss $\geq$2 mm	Subgingival plaque	Sterile paper point
Shiba et al., 2016 [49]	12	12 PI, 12 P	Japan	8.6 $\pm$ 7.2	PD $\geq$ 4 mm BOP and/or suppuration Radiographic bone loss	Subgingival plaque	Sterile paper point
Tsigarida et al., 2015 [50]	80	40 H, 20 PM, 20 PI	USA	$\geq$4 years	Clinical inflammation (redness, swelling, BOP, suppuration) Radiographic bone loss > 2 mm	Subgingival plaque	Sterile paper point
Jakobi et al., 2015 [51]	18	9 H, 9 PI, 9 P	Germany	>6 months	Presence of mobility BOP $\pm$ suppuration	Subgingival plaque	Sterile paper point
Zheng et al., 2014 [52]	24	10 H, 8 PM, 6 PI	China	Not stated	Zitzmann & Berglundh (2008)	Subgingival plaque	Periodontal probe
Schaumann et al., 2014 [53]	7	4.7 $\pm$ 3.6 PI	Germany	$\geq$1 year	PD $\geq$ 4 mm BOP Radiographic bone loss	Supra- and subgingival plaque	Sterile paper point
Maruyama et al., 2014 [54]	20	20 PI, 20 P	Japan	$\geq$1 year	PD $\geq$ 4 mm BOP $\pm$ suppuration Presence of radiographic bone loss	Subgingival plaque	Sterile paper point
Tamura et al., 2013 [55]	30	15 H, 15 PI	Japan	>6 months	PD $\geq$ 4 mm BOP and suppuration Radiographic bone loss	Subgingival plaque	Sterile paper point
Koyanagi et al., 2013 [21]	6	6 PI	Japan	Not stated	PD $\geq$ 5 mm BOP and/or suppuration Radiographic bone loss of more than three threads up to half of the implant length	Subgingival plaque	Sterile paper point

Table 3. *Cont.*

Author, Year	Number of Subjects	Number of Implants	Study Setting	Duration of Implant	Case Definition for Peri-Implantitis/Peri-Implant Mucositis	Samples Collected	Collection Method
Dabdoub et al., 2013 [25]	81	33 H, 20 PM, 20 PI	USA	≥1 year	Consensus Report of the Sixth European Workshop on Periodontology	Subgingival plaque	Sterile paper point
da Silva et al., 2013 [56]	20	10 PI, 20 H	Brazil	Not stated	PD ≥ 5 mm BOP and/or suppuration Saucer-shaped osseous defects of >3 mm	Subgingival plaque	Sterile Gracey curette
Kumar et al., 2012 [22]	40	10 H, 10 PI	USA	≥1 year	Classification of Periodontal Diseases (Armitage 1999) Consensus Report on Peri-Implant Diseases (Lindhe & Meyle 2008)	Subgingival plaque	Sterile paper point
Koyanagi et al., 2010 [57]	3	3 H, 3 PI	Japan	3–10	PD ≥ 5 mm BOP and/or suppuration Radiographic bone loss of more than three threads up to half of the implant length	Subgingival plaque	Sterile paper point
Faveri et al., 2010 [58]	50	25 H, 25 PI	Brazil	Not stated	PD ≥ 5 mm Saucer-shaped osseous defects of >3 mm BOP and/or suppuration	Subgingival plaque	Sterile Gracey curette

PD: probing depth; BOP: bleeding on probing; P: periodontitis; PI: peri-implantitis; H: healthy implant; PM: peri-implant mucositis.

Table 4. Summary of techniques of DNA extraction, amplification, and sequencing.

Author, Year	Method of DNA Extraction	DNA Amplification and Targeted Region	Sequencing Technique	Reference Database
Kim et al., 2023 [32]	Lucigen DNA kit, LGC Biosearch Technologies, Middleton, USA	PCR amplification of the 16s rRNA gene at the V3–V4 region	Illumina MiSeq	Human Oral Microbiome Database
Song et al., 2022 [33]	TIANamp Micro DNA Isolation Kit, TIANGEN BIOTECH, Beijing, China	PCR amplification at the V3–V4 hypervariable region of 16S rRNA with the primers 338F and 806R	Illumina MiSeq	Human Oral Microbiome database
Pallos et al., 2022 [34]	NucliSENS easyMAG, bioMérieux, Missouri, USA	V4 hypervariable region of the 16S rRNA gene was amplified using F515 and R80	Ion 318™ Chip kit v2 400-base chemistry	HOMD and Greengene amd NCBI 16s rRNA reference sequence
Barbagallo et al., 2022 [35]	PureLink Genomic DNA kit, Thermo Fisher Scientific, USA	PCR amplification of the 16s rRNA gene at V3–V4 region	Illumina Miseq	Human Oral Microbiome database
Shi et al., 2021 [36]	DNeasy PowerSoil kit, QIAGEN, Venlo, The Netherlands	PCR amplification of the 16S rRNA genes at V3–V4 region	Illumina MiSeq	Silva database
Polymeri et al., 2021 [37]	AGOWA mag Mini DNA Isolation Kit, LGC Genomics, Teddington, United Kingdom	PCR amplification of the 16S rRNA gene hypervariable region V5–V7.	454 GS-FLX + Titanium system was used for pyrosequencing	Ribosomal Database Project & Human Oral Microbiome Database
Korsch et al., 2021 [38]	Qiagen DNA MiniAmp Kit, QIAGEN, Venlo, The Netherlands	PCR amplification of the 16s rRNA gene at V1–V2 region	Illumina MiSeq	Silva database
Komatsu et al., 2020 [39]	Mora-extract, AMR Inc., Tokyo, Japan	Not stated	Illumina Miseq	Human Oral Microbiome database
Ghensi et al., 2020 [40]	Qiagen DNA MiniAmp kit, QIAGEN, Venlo, The Netherlands	Not stated	Illumina Hiseq	MetaPhlAn 2 and HUMAnN2
Aleksandrowicz et al., 2020 [41]	Genomic Mini kit, A&A Biotechnology, Gdańsk, Poland	The 2720 Thermal Cycler was used for the amplification of archaeal and bacterial DNA. Oligonucleotide-specific primers were used to target the specific 16s rRNA gene	3130xl Genetic Analyzer	GenBank

Table 4. *Cont.*

Author, Year	Method of DNA Extraction	DNA Amplification and Targeted Region	Sequencing Technique	Reference Database
Yu et al., 2019 [42]	Qiagen DNA MiniAmp kit, QIAGEN, Venlo, The Netherlands	PCR amplification at the hypervariable region V3–V4 of 16s rRNA	Paired-end MiSeq sequencing	Human Oral Microbiome Database
Kröger et al., 2018 [43]	Sigma-Aldrich GenElute Bacterial Genomic DNA Kit, Sigma-Aldrich, Munich, Germany	PCR amplification of the 16s rRNA gene at V3–V4 regions	Illumina MiSeq	Human Oral Microbiome Database
Gao et al., 2018 [44]	Not stated	PCR amplification of the 16S V3–V4 regions with primers 343F and 798R	Illumina Miseq	Human Oral Microbiome database
Daubert et al., 2018 [45]	Chelex-100, Bio-Rad, Hercules, USA	PCR amplification was used to amplify prokaryotic 16S rRNA genes using universal primers (27F and 1392R). Region of amplification not stated	Roche 454	Human Oral Microbiome database
Al-Ahmad et al., 2018 [46]	DNeasy Blood and Tissue kit, QIAGEN, Venlo, The Netherlands	PCR amplification of 16s rRNA using the universal primers 27F-YM and 1492R, region not stated	Ridom TraceEdit software, version 1.1.0	GenBank
Sousa et al., 2016 [47]	Not stated	Amplification with PCR using the 16S rRNA gene with V5–V7 primers	Illumina MiSeq	Greengenes
Sanz-Martin et al., 2017 [20]	Masterpure purification kit, Epicentre, Wisconsin, USA	PCR amplification of the 16s rRNA gene at V3–V4 region	Illumina MiSeq	Ribosomal Database Project (RDP)
Apatzidou et al., 2017 [23]	Proteinase K (100 mcg/mL) at 60 °C for 60 min, later boiled for 10 min Concentration measured with the Nanodrop NP-1000 spectrophotometer (Thermo Fisher Scientific, Renfrew, UK) Final concentration adjusted to 5 ng/mcL	PCR amplification of the V3–V4 region of the 16s rRNA gene	Illumina MiSeq	Greengenes database
Yu et al., 2016 [48]	Qiagen DNA MiniAmp kit, QIAGEN, Venlo, The Netherlands	PCR amplification of 16s rRNA at ca. 650 bp regions corresponding to the V2–V5 region	M13 forward primer	Human Oral Microbiome Database

Table 4. *Cont.*

Author, Year	Method of DNA Extraction	DNA Amplification and Targeted Region	Sequencing Technique	Reference Database
Shiba et al., 2016 [49]	Not stated	PCR amplification of 16s rRNA, region not stated	Illumina MiSeq	Human Oral Microbiome Database
Tsigarida et al., 2015 [50]	Qiagen DNA MiniAmp kit, QIAGEN, Venlo, The Netherlands	PCR amplification of the V1 to V3 and V7 to V9 regions	The TTitanium platform was used to perform multiplexed bacterial-tag-encoded FLX amplicon pyrosequencing.	Human Oral Microbiome Database
Jakobi et al., 2015 [51]	Qiagen DNA MiniAmp kit, QIAGEN, Venlo, The Netherlands	PCR amplification of 16s rDNA	Not stated	Ribosomal Database Project
Zheng et al., 2014 [52]	Not stated	PCR was used to amplify the V1–V3 regions of the 16s rRNA gene	The 454-GS-FLX sequencing platform was used for pyrosequencing	Ribosomal Database Project
Schaumann et al., 2014 [53]	QIAamp DNA MiniAmp Kit, QIAGEN, Venlo, The Netherlands	PCR amplification of 16s rRNA at the V1–V3 regions	Pyrosequencing was performed via the GS FLX sequencer	Greengenes
Maruyama et al., 2014 [54]	Mora-extract, AMR Inc. Tokyo, Japan	PCR amplification of the 16S V3–V4 regions with primers 806R and 515F	Roche 454	Ribosomal Database Project, Human Oral Microbiome Database, and NCBI
Tamura et al., 2013 [55]	Not stated	PCR amplification of the 16s rRNA gene with the forward primers 16S27F and 16S341F and the reverse primers 16S1492R and 16S907R	Takara Bio	GenBank database
Koyanagi et al., 2013 [21]	Mora-extract, AMR Inc. Tokyo, Japan	PCR amplification of the 16s rRNA gene with the primers 27F and 1492R	The 27F and 520R primers (BigDye Terminator Cycle Sequencing kit) were used, and 3130xl Genetic Analyzer	Ribosomal Database Project-II (RDP-II)
Dabdoub et al., 2013 [25]	Qiagen DNA MiniAmp kit, QIAGEN, Venlo, The Netherlands	PCR amplification of the 16s rRNA gene at two regions: V1–V3 and V7–V9	Pyrotag sequencing was performed	Greengenes

Table 4. *Cont.*

Author, Year	Method of DNA Extraction	DNA Amplification and Targeted Region	Sequencing Technique	Reference Database
da Silva et al., 2013 [56]	Masterpure DNA purification kit, Epicentre, Wisconsin, USA	Two step PCR was performed. The first step involved two sets of forward primers in a 1:1 ratio and the reverse primer 1541R. The second step involved the same two sets of forward primers and the reverse primer 1492R.	ABI Prism fluorescent bases	Ribosomal Data Project (RDP) & GenBank
Kumar et al., 2012 [22]	Qiagen DNA MiniAmp kit, QIAGEN, Venlo, The Netherlands	PCR amplification of 16s rRNA at the V1–V3 and V7–V9 regions	The Titanium platform was used to perform multiplexed bacterial-tag-encoded FLX amplicon pyrosequencing.	Greengenes
Koyanagi et al., 2010 [57]	Mora-extract, AMR Inc. Tokyo, Japan	PCR amplification of plasmid DNA	27F and 520R primers (BigDye Terminator Cycle Sequencing kit) were used and the 3130xl Genetic Analyzer	Ribosomal Database Project-II (RDP-II)
Faveri et al., 2010 [58]	Proteinase K (200 mg/mL) was added to the buffer and then inactivated at 95 °C	PCR amplification with the universal primer pair for Euryarchaea and the reverse primer 954rEyAr	ABI Prism fluorescent bases	Ribosomal Data Project (RDP) & GenBank

PCR: Polymerase chain reaction.

Table 5. Microbial profiles from the retrieved studies showing the diversity and richness and the abundance of taxa.

Author, Year	Groups	Results	
		Diversity and Richness	**Abundance of Taxa**
Kim et al., 2023 [32]	Peri-implantitis Periodontitis	PI = P [a] PI > P [b]	PI&P: *P. gingivalis*, *Prevotella* spp., *Treponema* spp., *F. alocis*, and *F. fastidiosum* PI > P: *Anaerotignum lactatifermentans*, *Bacteroides vulgatus*, *Faecalibacterium prausnitzii*, *Olsenella uli*, *Parasutterella excrementihominis*, *Prevotella buccae*, *P. alactolyticus*, and *Slackia exigua*
Song et al., 2022 [33]	Peri-implantitis	PI = HI [b] PI > HI [e] HI ≠ PI [c] (Significant difference between groups)	PI: *Bacteroidetes, Spirochaetes, and Synergistetes, as well as the genera of Porphyromonas,* *Treponema*, *Filifactor*, *Fretibacterium*, *Lachnospiraceae G-8*, and *Peptostreptococcaceae XIG-1* HI: *Proteobacteria*, *Neisseria*, *Streptococcus*, *Haemophilus*, and *Rothia*
Pallos et al., 2022 [34]	Peri-implantitis	HI > PI [a,e] HI = PI [c]	PI > HI: *Stenotrophomonas*, *Enterococcus*, *Leuconostoc genus*, *Faecalibacterium prausnitzii*, *Haemophilus parainfluenzae*, *Prevotella copri*, *Bacteroides vulgatus*, and *Bacteroides stercoris*
Barbagallo et al., 2022 [35]	Peri-implantitis Periodontitis	PI > P [a] PI = P [b]	PI: *Peptostreptococcaceae*, *Dialister*, *Mongibacterium*, *Atopobium*, and *Filifactor* P: *Bacteroidales*
Shi et al., 2021 [36]	Peri-implantitis Peri-implant mucositis	PI = PM (No significant difference between groups) [a,b,c]	PI = PM: No significant difference, *Bacteroidetes* (45.08% in PM, 42.89% in PI), *Firmicutes* (21.03% in PM, 19.44% in PI), *Proteobacteria* (11.16% in PM, 10.41% in PI) *Fusobacteria* (11.12% in PM, 14.7% in PI), *Spirochetes* (8.38% in PM, 9.68% in PI), *Porphyromonas* (17.04% in PM, 16.54% in PI), *Fusobacterium* (9.78% in PM, 12.39% in PI), *Treponema* (8.37% in PM, 9.59% in PI) and *Prevotella* (7.43% in PM, 7.04% in PI). PI > PM: *Holdemanella* and *Cardiobacterium* PM > PI: *Oribacterium*, *Staphylococcus*, and *Ramlibacter*
Polymeri et al., 2021 [37]	Peri-implantitis Peri-implant mucositis	HI = PM = PI (No significant differences between groups) [a,b,g]	PI: *Fusobacterium nucleatum* and *Treponema denticola* PM: *Rothia mucilaginosa* and *Streptococcus salivarius*
Korsch et al., 2021 [38]	Peri-implantitis	PI > HI [d]	PI: *Fusobacterium nucleatum* and *Porphyromonas gingivalis* HI: *Streptococcus*, *Neisseria*, *Rothia* and *Veillonella*
Komatsu et al., 2020 [39]	Peri-implantitis Periodontitis	PI > P [a] PI = P [c,g]	PI: *Solobacterium moorei* and *Prevotella denticola* P: *F. nucleatum*, *P. stomatis* and *Leptotrichia* sp.

Table 5. *Cont.*

Author, Year	Groups	Results	
		Diversity and Richness	Abundance of Taxa
Ghensi et al., 2020 [40]	Peri-implantitis Peri-implant mucositis	PI < HI [a,b]	PI: *Treponema maltophilum*, *Fretibacterium fastidiosum*, *Pseudoramibacter alactolyticus*, *T. lecithinolyticum*, *P. gingivalis*, *T. forsythia*, *Treponema denticola*, *P. endodontalis*, *Filifactor alocis*, and *Desulfobulbus* spp. HI: *C. gingivalis*, *C. granulosa*, *C. ochracea*, *S. noxia*, *S. artemidis*, *Actinomyces*, *Capnocytophaga*, *Neisseria*, *Rothia*, and *Streptococcus*
Aleksandrowicz et al., 2020 [41]	Peri-implantitis Periodontitis	Nil	PI: *F nucleatum* and *T denticola*
Yu et al., 2019 [42]	Peri-implantitis Periodontitis	PI = HI (No significant difference between groups) [d,f]	PI=HI: *Streptococcus infantis/mitis/oralis* (*HMT-070/HMT-071/HMT-638/HMT-677*) and *Fusobacterium* sp. *HMT-203/HMT-698* PI (Low abundance): *Aquificae*, *Chlamydiae*, *Gemmatimonadetes*, *Nitrospirae*, *TM6*, *Verrucomicrobia*, and *WPS2* phyla
Kröger et al., 2018 [43]	Peri-implantitis	PI > HI [g]	PI: *Eubacteriaceae* [XV], *Fretibacterium* sp. *HMT 362*, *Fretibacterium fastidiosum*, *Peptostreptococcaceae* [XI][G-6], *Alloprevotella* sp. *HMT 473*, *Fastidiosipila sanguinis*, *Filifactor alocis*, *Peptostreptococcaceae* [XI][G-4], *Bacteriodetes* [G-3] bacterium *HMT 365*, *Treponema parvum*, *Clostridiales* [F-1][G-1] bacterium *HMT 093*, and *Orobacterium*
Gao et al., 2018 [44]	Peri-implantitis	PI > HI [b] HI $\neq$ PI (Significant difference between groups) [c]	PI: *Moraxella*, *Micrococcus*, and *Acinetobacter* HI: *Neisseria*, *Haemophilus*, *Prevotella*, *Streptococcus*, *Porphyromonas*, *Clostridium*, *Capnocytophaga*, *Leptothrix*, *Actinomycetes*, and *Actinomyces*
Daubert et al., 2018 [45]	Peri-implantitis	HI > PI [a,b,c]	PI: *Veillonella* and *Neisseria.*
Al-Ahmad et al., 2018 [46]	Peri-implantitis	Not reported	PI: *Bacteroidetes* (phylum), *Fusobacterium nucleatum*
Sousa et al., 2016 [47]	Peri-implantitis Aggressive periodontitis Peri-implant mucositis	P > PI [a,b,f]	PI: *Propionibacterium*, *Paludibacterium*, *Staphylococcus*, *Filifactor*, *Mogibacterium*, *Bradyrhizobium*, and *Acinetobacter*

Table 5. *Cont.*

Author, Year	Groups	Results	
		Diversity and Richness	Abundance of Taxa
Sanz-Martin et al., 2017 [20]	Peri-implantitis	PI > HI [c]	PI: *Bacteroides*, *Spirochetes*, and *Synergistetes*, *Tannerella forsythia*, *Treponema denticola*, and *Porphyromonas gingivalis*, *Filifactor alocis*, *Fretibacterium fastidiosum*, and *Treponema maltophilum* HI: *Proteobacteria* and *Actinobacteria* PI > HI: *Porphyromonas* (phylum *Bacteroidetes*), *Treponema* (phylum *Spirochetes*), *Filifactor* (phylum *Firmicutes*), *Fretibacterium* (phylum *Synergistetes*), *Tannerella* (phylum *Bacteroidetes*), *T. forsythia*, *P. gingivalis*, and *T. denticola*). HI > PI: *Streptococcus* (phylum *Firmicutes*), *Veillonella* (phylum *Firmicutes*), *Rothia* (phylum *Actinobacteria*), *Haemophilus* (phylum *Proteobacteria*) and *Neisseria* spp.
Apatzidou et al., 2017 [23]	Peri-implantitis	PI > HI [a] HI = PI (No significant difference between groups) [b]	HI: *Actinobacillus* and *Streptococcus* PI: *Prevotella*, *Porphyromonas*, *Synergistetes*
Yu et al., 2016 [48]	Peri-implantitis Periodontitis	PI ≠ HI (Significant difference between groups) [f]	PI: High abundance of *F. fastidiosum* and *Fretibacterium*
Shiba et al., 2016 [49]	Peri-implantitis Smoking Periodontitis	PI = P (No significant difference between groups) [a,g] PI ≠ P (Significant difference between groups) [c]	PI = P: High rc-rRNA abundances *Porphyromonas gingivalis*, *Treponema denticola*, and *Tannerella forsythia*
Tsigarida et al., 2015 [50]	Peri-implantitis Smoking Peri-implant mucositis	HI = PI [b] HI ≠ PI (Significant difference between groups) [c]	PI: *Aggregatibacter*, *Capnocytophaga*, *Corynebacterium mucifaciens*, *Fretibacterium*, *Lachnoanaerobaculum*, *Lactobacillus panis*, *Neisseria*, *Prevotella* HI: *Actinomyces*, *Alloprevotella*, *Capnocytophaga*, *Enterobacter cancerogenus*, *Fusobacterium gonidiaformans*, *Fusobacterium*, *Lactobacillus johnsonii*, *Neisseria lactamica*, *Porphyromonas asaccharolytica*, *Prevotella enoeca*, *Prevotella*, *Pseudomonas*, *Pseudomonas pseudoalcaligenes*, *SR1 [G-1]*, *Streptococcus*, *Tannerella*
Jakobi et al., 2015 [51]	Peri-implantitis Periodontitis	Not reported	PI and P: *Enterococcus*, *Streptococcus*, *Porphyromonas*, *Fusobacterium*, *Prevotella*, *Bacillus*, and *Fretibacterium* Exclusive to PI: *Neisseria* and *Kingella* Exclusive to P: *Tannerella*, *Rothia*, *Parabacteroides*, *Parvimonas*, and *Filifactor* HI: *Enterococcus*, *Bacillus*, *Streptococcus*, *Fusobacterium*, *Prevotella*, *Porphyromonas*, *Rothia* and *Proteus*

Table 5. *Cont.*

Author, Year	Groups	Results	
		Diversity and Richness	Abundance of Taxa
Zheng et al., 2014 [52]	Peri-implantitis Peri-implant mucositis	PM = PI (No significant differences among groups) [f] HI > PM [f] HI > PI [f] PI > HI [a,b,g]	PI: *Leptotrichia hofstadii, Eubacterium infirmum, Kingella denitrificans, Actinomyces cardiffensis, Eubacterium minutum, Treponema lecithinolyticum*, and *Gemella* sanguinis, *Gemella sanguinis, Eubacterium minutum*, and *Actinomyces cardiffensis*
Schaumann et al., 2014 [53]	Peri-implantitis Periodontitis	PI = P (No significant difference between groups) [a]	PI: *Porphyromonadaceae, Lachnospiraceae*, and *Streptococcaceae;* Genera *Rothia, Actinomyces, Paenibacillus, Microbacterium, Pseudoramibacter, Leptotrichia, Parascardovia, Tannerella, Granulicatella, Tessaracoccus, Clostridium, Aeromonadales, Veillonella, Capnocytophaga, Prevotella, TG5, Fusobacterium, Exiguobacterium, Enterococcus, Porphyromonas* and *Streptococcus*.
Maruyama et al., 2014 [54]	Peri-implantitis Periodontitis	PI = P [a,b,c,g] (no significant difference)	PI: *Prevotella nigrescens, Olsenella, Sphingomonas, Peptostreptococcus*, and *Neisseriaceae* P: *Peptostreptococcaceae* sp. and *Desulfomicrobium orale*
Tamura et al., 2013 [55]	Peri-implantitis	Not reported	PI: *E nodatum, P intermedia, F nucleatum, Filifactor alocis, E brachy, Parascardovia denticolens, Parvimonas micra* HI: *Veillonella* sp., *Propionibacterium acnes, Pseudoramibacter alactolyticus, Parvimonas micra*
Koyanagi et al., 2013 [21]	Peri-implantitis Periodontitis	PI > P [a,b]	PI and P: *Firmicutes* and *Bacteroidetes, Fusobacterium* spp. and *Streptococcus* spp., Exclusive to PI: *Parvimonas micra, Peptostreptococcus stomatis, Pseudoramibacter alactolyticus*, and *Solobacterium moorei* PI > P sites: *Dialister* spp., *Eubacterium* spp., *Porphyromonas* spp., *P. gingivalis*. PI = P sites: *T. forsythia, T. denticola*
Dabdoub et al., 2013 [25]	Peri-implantitis Periodontitis	P > PI [a]	PI = P: No significant difference in the number of shared species
da Silva et al., 2013 [56]	Peri-implantitis	Not reported	HI: *Actinomyces, Atopobium, Gemella, Kingella* and *Rothia, Campylobacter, Desulfobulbus, Dialister, Eubacterium, Filifactor, Mitsukella, Porphyromonas* and *Pseudoramibacter*. PI > HI: *Fusobacterium nucleatum, Dialister invisus, Streptococcus* sp. human oral taxon (*HOT*) *064, Filifactor alocis*, and *Mitsuokella* sp. *HOT 131* HI > PI: *Veillonella dispar, Actinomyces meyeri*, and *Granulicatella adiacens*

Table 5. *Cont.*

Author, Year	Groups	Results	
		Diversity and Richness	**Abundance of Taxa**
Kumar et al., 2012 [22]	Peri-implantitis Periodontitis	HI > PI ^{c} P > PI ^{a}	PI: *Actinomyces*, *Peptococcus*, *Campylobacter*, *nonmutans Streptococcus*, *Butyrivibrio*, and *Streptococcus mutans*, *B. fibrisolvens*
Koyanagi et al., 2010 [57]	Peri-implantitis Periodontitis	PI > P ^{a,b}	PI: *Chloroflexi*, *Tenericutes*, and *Synergistetes* phyla Exclusive to PI: *Parvimonas* micra, *Peptostreptococcus stomatis*, *Pseudoramibacter alactolyticus*, *Fusobacterium nucleatum*, and *Solobacterium moorei* Detected in P: *Fusobacterium nucleatum*, *Granulicatella adiacens*
Faveri et al., 2010 [58]	Peri-implantitis	Not reported	PI: *Archaea* detected at significantly higher abundance

PI: Peri-implantitis; HI: healthy implants; P: periodontitis; PM: peri-mucositis. ^{a}: Shannon's index; ^{b}: Chao1 index; ^{c}: Principal Coordinate Analysis (PCoA); ^{d}: permutational multivariate analysis of variance (PERMANOVA); ^{e}: InvSimpson's index; ^{f}: weighted Unifrac distance analysis; ^{g}: number of operational taxonomic units (OTUs).

2.2. Microbial Profile

Koyanagi et al. revealed that implants with peri-implantitis had a higher abundance of *Eubacterium* spp. when compared to healthy implants, and this finding is also supported by Zheng et al. and Kroger et al. [21,43,52]; da Silva et al. found that healthy implants demonstrated lower proportions of *Eubacterium* compared to peri-implantitis sites, while Koyanagi et al. and Zheng et al. concluded that peri-implantitis sites had significantly higher proportions of *Eubacterium* [21,52,56]. Sanz-Martin et al. reported higher levels of *Eubacterium* in a healthy implant, when a diseased implant was also present in the same oral cavity [20]. Two studies found high levels of *Bacteroidetes* and *Firmucutes* in PI sites as compared to HI sites [20,46]. Three authors found higher levels of *Bacteroides* in diseased implants [32–34]. Yu et al. demonstrated that *F. fastidiosum SH03* and the *Fretibacterium* oral taxon *SH01* were linked with plaque at healthy subgingival sites [48]. This study concluded that there were no clear differences or similarities between *Synergistetes* communities found in diseased versus healthy sites or between periodontal/subgingival niches and peri-implant/submucosal niches [48]. Another study by Yu et al. also showed that the prevalent and abundant bacteria were *Streptococcus infantis/mitis/oralis (HMT-070/HMT-071/HMT-638/HMT-677)* and *Fusobacterium* sp. *HMT-203/HMT-698* in healthy implants and diseased implants [42]. Another 18 phyla were found in low abundance, particularly the *Aquificae, Chlamydiae, Gemmatimonadetes, Nitrospirae, TM6, Verrucomicrobia,* and *WPS2* phyla, which were present in <0.01% of the total reads for each of the four clinical site categories, with some being undetectable in one or more niches [42]. Healthy implants demonstrated higher proportions of *Actinomyces, Atopobium, Gemella, Kingella* and *Rothia* and lower levels of *Campylobacter, Desulfobulbus, Dialister, Eubacterium, Filifactor, Mitsukella, Porphyromonas,* and *Pseudoramibacter* in one study [56]. One study that underwent a pathogen-specific analysis for *Archaea* found that PI sites had a higher frequency of sites that were positive for *Archaea* [58]. *Filifactor* was found to be abundant in peri-implantitis sites when compared with healthy implant sites, as shown by several studies [20,35,36,40,47,55,56]. Three studies demonstrated that *Parvimonas* was the most abundant at peri-implantitis sites [21,55,57].

2.2.1. Phyla

The range of phyla was reported to be varied among the 25 studies. Koyanagi T et al. reported that *Firmicutes* (45.6%) is the most abundant phylum found in the subgingival plaque in peri-implantitis samples, followed by *Bacteroidetes, Proteobacteria, Fusobacteria, Actinobacteria,* TM7, *Synergistetes, Spirochaetes, Tenericutes, Chloroflexi,* and *Deferribacteres* [21]. Three studies were in concordance in concluding that *Bacteroidetes* is one of the genera that is found in great abundance in peri-implantitis samples [20,21,46]. The abundance of *Synergistetes* was reported to be higher in diseased samples in four studies in comparison to in healthy samples [20,21,23,33]. *Spirochaetes* was identified in diseased samples in three studies [20,21,46], with one study reporting that *Spirochaetes* increased significantly as peri-implantitis became more severe [20].

2.2.2. Genus

Numerous changes were reported at the genus level (Table 5), with many of them focusing on several genera which are the most abundant in the peri-implant sites. One study reported that there was a preponderance of *Veillonella* in diseased peri-implant mucosal tissues [45]. However, there are also studies that have suggested that *Veillonella* is significantly reduced in samples with an increasing peri-implantitis severity [20,53]. *Veillonella* was also associated with healthy implant sites in other studies [20,47,55,56]. Several authors have found that *Prevotella* spp. are significantly more abundant at peri-implantitis sites [23,34,36,39,53,54]. Kumar et al. and Daubert et al. found that healthy implants showed higher levels of these two microorganism species [22,45], which was also supported by Apatzidou et al., who showed their greater abundance in diseased samples [23]. Other than *Veillonella* and *Prevotella*, most studies also pointed out that *Porphyromonas* was commonly associated with diseased implants [20,23,51,53,56]. Several studies pointed out that *Fusobacterium* was present in high levels in peri-implantitis samples [21,37,41,46,55–57]. Five studies reported that *Streptococcus* was more abundant in healthy plaque samples as compared to its abundance in diseased samples [20,22,23,44,45]. Yu et al. also found that *Streptococcus* was found in both healthy implants and peri-implantitis sites [42]. On the contrary, Kumar et al. concluded that peri-implantitis samples demonstrated a higher level of *Streptococcus* [22]. A study reported that *Propionibacterium, Paludibacter, Staphylococcus, Filifactor, Mogibacterium, Bradyrhizobium,* and *Acinetobacter* are unique to peri-implant sites [47]. In addition, *Actinomyces* spp. has been reported to be prevalent in peri-implantitis sites [22,52,53]. However, da Silva et al. reported higher levels of *Actinomyces* spp. in healthy implants [56].

2.2.3. Microbiome Complex

Apart from the genera and phyla levels, Al-Ahmad et al. and Kim et al. reported that *Porphyromonas gingivalis* and *Tannerella forsythia* of the red complex are highly associated with peri-implantitis [32,46]. A study reported positive correlations with certain red and orange complex bacteria but a negatively correlation with blue complex bacteria in peri-implantitis samples [20]. Furthermore, another study reported that *Bacteroidetes, Chloroflexi, Spirochaetes, Synergistetes,* and *TM7* positively corresponded with the pocket depths [23].

2.2.4. Peri-Implantitis with Periodontitis

Granulicatella adiacens (phylum *Bacillota*) was identified in two-thirds of peri-implantitis sites; these two species were also detected at periodontitis sites but not in healthy implants [57]. Shiba et al. found that the microbial composition at the genus level was diverse among the samples for each disease and between both samples from each individual, although the predominant species were similar [49]. Two studies showed that the periodontitis microbial community is more diverse than peri-implantitis sites [25,47]. Interestingly, three studies found the opposite, whereby periodontitis samples yielded lower diversities than peri-implantitis samples [21,22,57]. Aleksandrowicz et al. demonstrated that *Archaea* was found in diseased implants and teeth [41]. Furthermore, they were found in abundant levels at periodontitis sites when compared to peri-implantitis sites [41].

2.2.5. Peri-Implantitis with Peri-Implant Mucositis

Shi et al. reported no differences in diversity between peri-mucositis sites as compared to peri-implantitis sites, but they found an increased microbial richness in peri-mucositis sites [36]. Sousa et al. reported a decreased abundance of *Bradyrhizobium* in peri-mucositis sites and peri-implantitis sites [47]. One study concluded that the microbial profile associated with peri-implantitis was also present with a moderate relative abundance at peri-mucositis sites. This study also found that the Shannon index of peri-mucositis was lower than that of peri-implantitis [52]. Tsigarida et al. reported subtle differences between the peri-mucositis and peri-implantitis microbiomes, and these subtle differences were between the transition from health to disease [50]. *Streptococci* and *Rothia* were associated with peri-mucositis, while *Fusobacterium* and *Treponema* were associated with peri-implantitis, as shown by Polymeri et al. [37]

2.3. Heterogeneity of Studies

Significant heterogeneity can be identified in the methodologies of the selected studies. The ROBINS-E tool was used to assess the quality of the 32 nonrandomized cohort observational studies. The ROBINS-E tool (Table 1) showed that nine studies had some concerns, while four studies were at a high risk of bias. Table 4 illustrates the heterogenicity of the gene sequencing techniques utilized. Figure 2 illustrates the diversity reported in terms of the Shannon's indexes reported by five studies [21,25,36,37,57]. Figure 3 illustrates the heterogeneity regarding the location (Figure 3a), database used (Figure 3b), and case definition criteria (Figure 3c) of the studies reviewed.

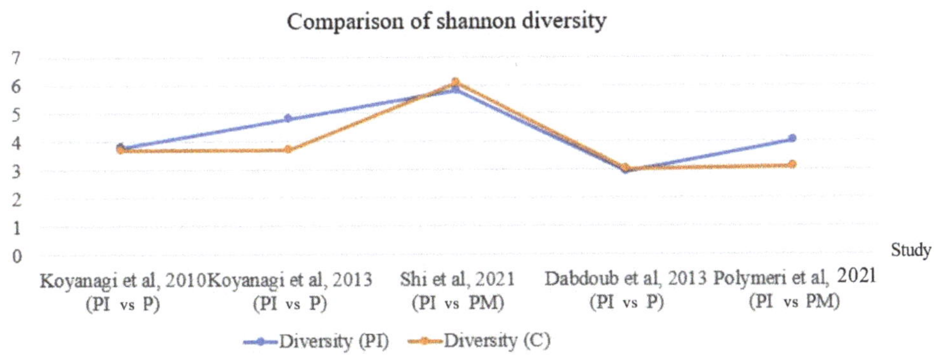

Figure 2. Different Shannon's indexes reported by the studies reviewed. PI—peri-implantitis, PM—peri-implant mucositis, P—periodontitis, C—comparison group [21,25,36,37,57].

(a)

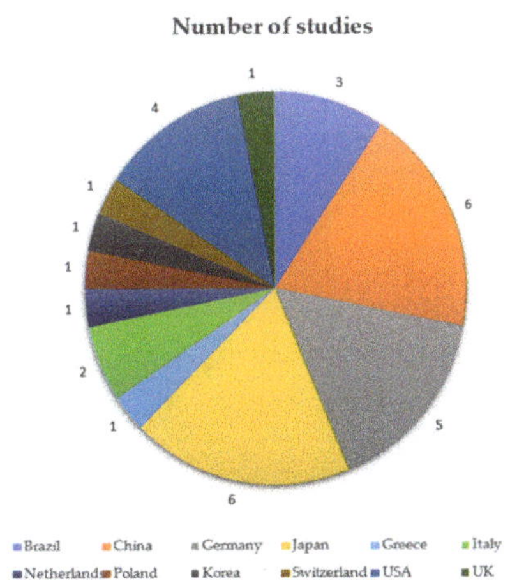

(b)

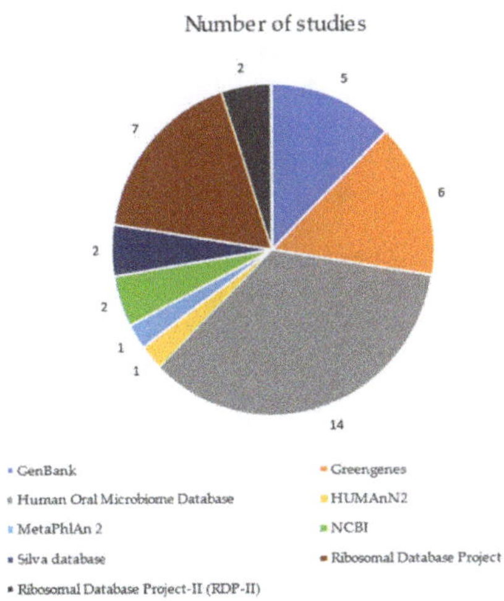

Figure 3. *Cont.*

Figure 3. (**a**) Location; (**b**) database used; (**c**) case definition criteria of the studies reviewed. PD: probing depth; BOP: bleeding on probing [20–23,25,32,34,36–50,52–58].

3. Discussion

This systematic review comprehensively reviews the current available evidence on the microbiome of peri-implantitis. Variations in the study methods, sample collection, and study design were observed. However, the review focuses on studies employing the 16S r RNA gene sequencing technique to summarize meaningful observations from the available evidence.

Ten of the studies reviewed showed that the microbial diversity of peri-implantitis is distinct and usually higher than that at healthy implant sites [14,15,17,19,24,26,28,34,38,39]. The alpha diversity considers the richness (number of taxa) and evenness (relative abundance) of species within a sample/community; the beta-diversity quantifies the identities of taxa involved between samples/communities [49]. Changes in oxygen and nutrient concentrations associated with the deepening of a pocket around an implant may be responsible for the shift in the microbial diversity [32]. Figure 2 shows the Shannon's indexes reported by five studies, as not all studies reported indices [21,25,36,37,57]. These variations in the diversity can be explained by the heterogenicity of various factors such as the location of the study (Figure 3a), the reference database (Figure 3b), and the case criteria definition (Figure 3c). A variation in the genomic database can introduce conflicting results, as one study showed that even the use of a single database within a study can implicate systematic errors during the mapping process which subsequently affects genomic analyses [59]. In addition to that, the sample collection method and the type of sample collected are other confounding factors that may produce conflicting findings.

The studies that included in the current review originate from different countries (Figure 3a), for example, Japan [21,49,55,57], China [36,42,48,52,60], United States of America [22,25,45,50], United Kingdom [47], Germany [38,43,46,53], and The Netherlands [37]. It is significant to note that certain sections of the globe are not represented here. This may also be due to the exclusion of articles written in other languages. Hence, the current data may be significantly influenced by the diet and genetic make-up of the individuals from the representative countries [61]. The characterization of oral dysbiosis in different ethnicities and races presents significant challenges due to variations across multiple studies [62–64]. This is due to the highly varied diet, nutrition and lifestyle practices present over several generations in different geographical locations [65,66].

The case definition for peri-implantitis varied across the studies reviewed (Figure 3c). For example, Koyanagi et al. used a criteria of a probing depth (PD) ≥5 mm with bleeding on probing (BOP) and/or suppuration and bone loss >3 threads up to half of the implant length, while Apatzidou et al. diagnosed subjects as having peri-implantitis when there was PD ≥ 6 mm, BOP and/or suppuration, and radiographic bone loss of ≥2 mm in at least one implant surface after one year of loading [21,23]. However, it is evident that the disease severity may vary, even with the employment of the above criteria, hence making it difficult to combine or compare the results of certain studies. Standardizing the methodological quality of microbiome studies has been previously suggested as a necessary step in this direction.

Even though few studies included criteria related to the systemic status of the patient, drugs taken, previous history of other oral diseases like periodontitis and the age of the patient into consideration, the varied criteria set across studies makes a meaningful comparison irrelevant. It would be greatly beneficial for future investigations into the microbiome of the oral cavity to follow a standardized protocol to facilitate comparability between studies [67].

The reviewed studies provide a deeper understanding of the microbial profile of peri-implantitis. However, the different DNA extraction kits used may have had an influence on the microbial data, for example, the Qiagen DNA MiniAmp kit, (QIAGEN, Venlo, The Netherlands) [22,25,38,42,48,50,53], GenElute Bacterial Genomic DNA kit, (Sigma-Aldrich, Munich, Germany) [43], Mora-extract kit, (AMR Inc., Tokyo, Japan) [21,57], Real-time PCR with TaqMan Probe, (Thermo Fisher Scientific, Waltham, MA, USA) [23], DNeasy Kit, (QIAGEN, Venlo, The Netherlands) [36,46], and the Masterpure purification kit, (Epicentre, Verona, Wisconsin, USA) [20,56].

Despite being considered an extension of peri-implantitis and the presence of common bacteria, peri-implant mucositis has been reported to have a distinct microbial profile in some studies [68,69]. However, a few studies were not able to provide a conclusive result on this aspect [36,37,47,50,52]. The diversity in peri-implant mucositis has been reported to be higher than at healthy implant sites [36] but lower than in peri-implantitis [52]. Moreover, the immune cell profiles of both entities seem to differ as well. Enhanced neutrophil and B-cell responses have previously been identified for peri-implantitis lesions when compared to peri-implant mucositis lesions under experimental conditions. The shift in the microbiome profile may also be explained by the increase in frequency and the number of bleeding sites subsequent to biofilm accumulation surrounding the implants [70].

The association of *Veillonella* sp. with healthy implant sites is well-correlated with its reduced prevalence in deeper pockets and severe disease progression [20,43,46,55]. *Streptococci* spp. have been identified in both diseased [21,22,53,56] and healthy sites [20,23,45]. *Neisseria* sp. have been associated with healthy implants and negatively correlates with the probing depth [20,40,43,44], suggesting that *Neisseria* sp. could have been replaced by other colonizers or may exert a protective effect. Species of the genus *Neisseria* are well-established primary colonizers of the dental plaque of natural teeth but are not well known for their presence in dental implants. On the contrary, three studies reported high levels of *Neisseria* sp. in peri-implantitis sites, which contradicts other studies [22,51,54]. Considering the common occurrence of these species in the oral cavity and the possibility of transfer from a diseased to a healthy site or vice versa leads to the lack of a clear understanding of its role in the initiation and the progression of the disease.

Numerous studies have identified *Fusobacterium* sp. as the dominant species in peri-implantitis [20,21,46]. Studies have also reported the presence of the genus *Treponema* at peri-implantitis sites of increasing severity [20,43]. However, Kumar et al. reported higher levels of the genera *Treponema* and *Prevotella* at healthy implant sites, which is the opposite to what other studies have found [22]. Peri-implantitis sites have also seen an abundance of species from the phylum *Synergistetes* [20,23,46]. *Porphyromonas* sp. have been reported at peri-implantitis sites by multiple studies [20,21,23].

A distinct microbial pattern could not be identified across all the 25 studies reviewed, possibly due to the abovementioned factors. Sahrmann et al. also found that there was an absence of a characteristic bacterial profile at peri-implantitis sites [71]. Both the current review and the review by Sahrmann et al. had a consensus that there was considerable heterogeneity in the studies reviewed [71]. The red complex is frequently identified at peri-implantitis sites, as are putative pathogens of the orange and yellow complex. Furthermore, it seems that the relative abundance of each complex changes with an increasing disease progression severity. The blue complex was also reported to be negatively correlated with peri-implantitis sites, suggesting its protective effect. The red complex was also more abundant at implant sites for subjects who smoked, which correlates well with our current understanding that smoking is a risk factor for peri-implantitis. The studies have findings that contradict one another, and this makes it difficult to obtain a characteristic microbial profile for peri-implantitis. However, it is evident that the microbiome of peri-implantitis is unique and distinct from that of periodontitis.

Carvalho et al. found that peri-implantitis lesions were associated with the presence of *S. epidermidis, P. gingivalis, T. forsythia, T. denticola, F. nucleatum*, and *P. intermedia* [72]. The review included culture-dependent studies in the analysis. On the contrary, the current systematic review only included studies that utilized next-generation sequencing due to its improved detection limit [30,73]. Additionally, Carvalho et al. reported that a definitive conclusion regarding the microbiome of peri-implantitis could not be reached due to the nature of the studies analyzed. Next-generation sequencing methods have shown that the microbiome of peri-implantitis is distinct from that of periodontitis. Non-culturable species such as *Fusobacterium* and the *Treponema* sp. *HMT-257* have been detected in peri-implantitis lesions [74,75]. The current systematic review demonstrates that, even with the inclusion of only next-generation sequencing studies, a distinct and unique microbial community pattern could not be identified.

The current review is limited by the studies' number of participants, with the highest being 139 in a study by Aleksandrowicz et al. [41]. This suggests that the results may not be generalized to the clinical setting due to the small sample size. This review is also limited by the heterogeneity presented across all studies reviewed. Hence, a characteristic microbial profile cannot be determined for future targeted therapies.

4. Materials and Methods

A systematic review of observational and case-control studies (PROSPERO) (CRD42023459266) investigating the microbiome of peri-implantitis lesions was performed on the Cochrane, Medline, and Scopus databases from inception until 1 August 2023 and reported according to the Preferred Reporting Items for Systematic Reviews and Meta-Analysis (PRISMA) [76]. A focused question was formulated based on PECO (population, exposure, comparator, and outcome). The population included patients with at least one osseointegrated dental implant, the exposure was the diagnosis of peri-implantitis lesions, the comparator included healthy implants, periodontitis sites, as well as peri-implant mucositis sites, and the outcome measure was the bacterial composition obtained from samples taken from peri-implantitis sites, as assessed through next-generation sequencing. The question was as follows: Among patients with at least one osseointegrated dental implant, what would be the difference between peri-implantitis lesions, healthy implants, periodontitis, and peri-implant mucositis in terms of the bacterial composition obtained from samples as assessed via next-generation sequencing?

The search strategy involved a combination of the following key terms: peri-implantitis, inflammation, disease, infection, consequence, sequence analysis, RNA, 16S, metagenomics, metagenome, microbiota, and bacteria. The keywords were combined using the Boolean operators "AND" and "OR" in the strategic search. This systematic review followed the Preferred Reporting Items for Systematic Reviews and Meta-Analyses (PRISMA) criteria [77].

The titles and abstracts were independently screened by two reviewers (K.C.G., R.K.M.) for eligible studies, followed by full-text reading. Data were extracted independently and in duplicate by the two reviewers (K.C.G., R.K.M.) into a data extraction form created following the Cochrane Handbook of Systematic Reviews of Interventions guidelines [76]. Observational and case-control studies investigating the microbiome of peri-implant tissues through next-generation DNA sequencing methods were included. Culture-based studies, conference papers, review articles, studies regarding peri-implantitis associated with other systematic factors (diabetes mellitus, immune disorders, etc.), and articles that examined only specific microorganisms were excluded from this systematic review. Non-English language articles and research conducted on non-human specimens were also excluded. This was followed by full-text screening for eligibility. The complete search strategy used is shown in Table 6. Table 7 depicts the inclusion and exclusion criteria for the articles.

Table 6. Search strategies employed.

Database	Search Terms
Medline	(Peri-implantiti$ OR Peri adj2 Implantiti$ OR Peri-implant$ adj2 inflam$ OR Peri-implant$ adj2 infect$ OR Peri-implant$ adj2 disease$ OR exp Peri-Implantitis/or exp Dental Implants/or exp Dental Implantation, Endosseous/OR peri-implant adj2 mucositi$ OR peri adj2 implant adj2 mucositi$ OR periimplant adj2 mucositi$ OR periimplant$ adj2 mucos$) AND (exp sequence analysis/or exp sequence analysis, dna/or exp sequence analysis, rna/or exp rna-seq/OR exp RNA, Ribosomal, 16S/OR exp Microbiota/OR exp Bacteria/)
Cochrane	(peri-implantiti* OR periimplantiti* OR (Peri-Implantitis):ti,ab,kw OR Peri-implant* NEAR/2 inflam* OR Peri-implant* NEAR/2 infect* OR peri-implant muco*sitis OR peri-implant NEAR/2 disease* OR peri-implant infect* OR MeSH descriptor: [Peri-Implantitis] explode all trees OR periimplant* NEAR/2 mucos*) AND (dental implant* OR dental implant, endosseous OR endosseous dental implant*) AND (MeSH descriptor: [Sequence Analysis, DNA] explode all trees OR MeSH descriptor: [Sequence Analysis] explode all trees OR MeSH descriptor: [Sequence Analysis, RNA] explode all trees OR MeSH descriptor: [RNA-Seq] explode all trees OR MeSH descriptor: [RNA, Ribosomal, 16S] explode all trees OR MeSH descriptor: [Microbiota] explode all trees OR MeSH descriptor: [Bacteria] explode all trees)
Scopus	(peri-implant* OR peri W/2 implant* OR peri-implant* W/2 inflam* OR peri-implant* W/2 infect* OR peri-implant* W/2 disease* OR peri-implant W/2 mucositi* OR peri W/2 implant W/2 mucositi* OR periimplant W/2 mucositi* OR periimplant* W/2 mucos*) AND (dental AND implants OR dental AND implantation AND endosseous) AND ((sequence AND analysis) OR (sequence AND analysis AND dna) OR (sequence AND analysis AND rna) OR rna-seq OR (rna AND ribosomal AND 16s)) AND (microbiota OR bacteria)

ti: Title; ab: Abstract; kw: Keywords; exp: Explode.

Table 7. Inclusion and exclusion criteria used for the studies screened.

Inclusion Criteria	Exclusion Criteria
Observational and case-control studies investigating the microbiome of peri-implant tissues through next-generation DNA sequencing methods. Human studies in English	Culture-based studies, conference papers, review articles, studies regarding peri-implantitis associated with other systematic factors (diabetes mellitus, immune disorders, etc.) Articles that examined only specific microorganisms. Non-English language articles and research conducted on non-human specimens.

The relevant studies were assessed with the Risk Of Bias In Non-randomized Studies-of Exposures (ROBINS-E) tool [78].

5. Conclusions

The study of the microbiome with next-generation sequencing allows more insight into the possible casual relationships between the bacteria and diseased state and not just culturable or cultivatable species. A unique and distinct microbial pattern could not be identified due to the vast heterogeneity present across all studies. The authors propose that future studies should investigate the microbial profile of peri-implantitis based on the severity of the disease to further provide insight into the progression and alteration of the microbial community within the peri-implant pocket.

A universal clinical parameter for the diagnosis of peri-implantitis should be implemented in all future studies to minimize the confounding factors. The subject pool should also be more diverse and larger to compensate for individual differences, and perhaps, a distinct microbial profile may be seen with a larger sample size. The studies reviewed also show that different groups of bacteria exist in the pockets at different stages of the diseases. This may imply that, with a complete microbial profile, an accurate estimation of the disease progression and monitoring can be performed. Furthermore, this also allows targeted drug therapies towards selective microorganisms that are strongly associated with peri-implantitis.

Author Contributions: Conceptualization, R.K.M.; methodology, R.K.M. and K.C.G.; formal analysis, R.K.M. and K.C.G.; data curation, R.K.M. and K.C.G.; writing—original draft preparation, K.C.G.; writing—review and editing, R.K.M.; supervision, R.K.M. All authors have read and agreed to the published version of the manuscript.

Funding: The APC was funded by Ajman University.

Institutional Review Board Statement: Not applicable.

Informed Consent Statement: Not applicable.

Data Availability Statement: All data are provided with the manuscript.

Conflicts of Interest: The authors declare no conflict of interest.

References

1. Buser, D.; Janner, S.F.M.; Wittneben, J.-G.; Brägger, U.; Ramseier, C.A.; Salvi, G.E. 10-Year Survival and Success Rates of 511 Titanium Implants with a Sandblasted and Acid-Etched Surface: A Retrospective Study in 303 Partially Edentulous Patients: 10-Year Survival and Success Rates of SLA Implants. *Clin. Implant Dent. Relat. Res.* **2012**, *14*, 839–851. [CrossRef] [PubMed]
2. Serino, G.; Hultin, K. Periimplant Disease and Prosthetic Risk Indicators: A Literature Review. *Implant Dent.* **2019**, *28*, 125–137. [CrossRef] [PubMed]
3. Pesce, P.; Canullo, L.; Grusovin, M.G.; de Bruyn, H.; Cosyn, J.; Pera, P. Systematic Review of Some Prosthetic Risk Factors for Periimplantitis. *J. Prosthet. Dent.* **2015**, *114*, 346–350. [CrossRef] [PubMed]
4. Derks, J.; Tomasi, C. Peri-Implant Health and Disease. A Systematic Review of Current Epidemiology. *J. Clin. Periodontol.* **2015**, *42*, S158–S171. [CrossRef]

5. Rodrigo, D.; Sanz-Sánchez, I.; Figuero, E.; Llodrá, J.C.; Bravo, M.; Caffesse, R.G.; Vallcorba, N.; Guerrero, A.; Herrera, D. Prevalence and Risk Indicators of Peri-Implant Diseases in Spain. *J. Clin. Periodontol.* **2018**, *45*, 1510–1520. [CrossRef]
6. Roos-Jansåker, A.-M.; Lindahl, C.; Renvert, H.; Renvert, S. Nine- to Fourteen-Year Follow-up of Implant Treatment. Part II: Presence of Peri-Implant Lesions. *J. Clin. Periodontol.* **2006**, *33*, 290–295. [CrossRef]
7. Fransson, C.; Lekholm, U.; Jemt, T.; Berglundh, T. Prevalence of Subjects with Progressive Bone Loss at Implants: Prevalence of Subjects with Progressive Bone Loss at Implants. *Clin. Oral Implant Res.* **2005**, *16*, 440–446. [CrossRef] [PubMed]
8. Diaz, P.; Gonzalo, E.; Villagra, L.J.G.; Miegimolle, B.; Suarez, M.J. What Is the Prevalence of Peri-Implantitis? A Systematic Review and Meta-Analysis. *BMC Oral Health* **2022**, *22*, 449. [CrossRef] [PubMed]
9. Astolfi, V.; Ríos-Carrasco, B.; Gil-Mur, F.J.; Ríos-Santos, J.V.; Bullón, B.; Herrero-Climent, M.; Bullón, P. Incidence of Peri-Implantitis and Relationship with Different Conditions: A Retrospective Study. *Int. J. Environ. Res. Public Health* **2022**, *19*, 4147. [CrossRef] [PubMed]
10. Hashim, D.; Cionca, N. A Comprehensive Review of Peri-Implantitis Risk Factors. *Curr. Oral Health Rep.* **2020**, *7*, 262–273. [CrossRef]
11. Kumar, P.S. Systemic Risk Factors for the Development of Periimplant Diseases. *Implant Dent.* **2019**, *28*, 115–119. [CrossRef]
12. Steiger-Ronay, V.; Merlini, A.; Wiedemeier, D.B.; Schmidlin, P.R.; Attin, T.; Sahrmann, P. Location of Unaccessible Implant Surface Areas during Debridement in Simulated Peri-Implantitis Therapy. *BMC Oral Health* **2017**, *17*, 137. [CrossRef]
13. Apatzidou, D.A. Modern Approaches to Non-Surgical Biofilm Management. In *Frontiers of Oral Biology*; Kinane, D.F., Mombelli, A., Eds.; KARGER: Basel, Switzerland, 2011; Volume 15, pp. 99–116. ISBN 978-3-8055-9834-7.
14. James, P.; Worthington, H.V.; Parnell, C.; Harding, M.; Lamont, T.; Cheung, A.; Whelton, H.; Riley, P. Chlorhexidine Mouthrinse as an Adjunctive Treatment for Gingival Health. *Cochrane Database Syst. Rev.* **2017**, *2021*, CD008676. [CrossRef]
15. Lombardo, G.; Signoriello, A.; Corrocher, G.; Signoretto, C.; Burlacchini, G.; Pardo, A.; Nocini, P.F. A Topical Desiccant Agent in Association with Manual Debridement in the Initial Treatment of Peri-Implant Mucositis: A Clinical and Microbiological Pilot Study. *Antibiotics* **2019**, *8*, 82. [CrossRef]
16. Herrera, D.; Berglundh, T.; Schwarz, F.; Chapple, I.; Jepsen, S.; Sculean, A.; Kebschull, M.; Papapanou, P.N.; Tonetti, M.S.; Sanz, M.; et al. Prevention and Treatment of Peri-Implant Diseases—The EFP S3 Level Clinical Practice Guideline. *J. Clin. Periodontol.* **2023**, *50*, 4–76. [CrossRef] [PubMed]
17. Ferreira, S.D.; Martins, C.C.; Amaral, S.A.; Vieira, T.R.; Albuquerque, B.N.; Cota, L.O.M.; Esteves Lima, R.P.; Costa, F.O. Periodontitis as a Risk Factor for Peri-Implantitis: Systematic Review and Meta-Analysis of Observational Studies. *J. Dent.* **2018**, *79*, 1–10. [CrossRef] [PubMed]
18. Renvert, S.; Lindahl, C.; Renvert, H.; Persson, G.R. Clinical and Microbiological Analysis of Subjects Treated with Brånemark or AstraTech Implants: A 7-Year Follow-up Study. *Clin. Oral Implant Res.* **2008**, *19*, 342–347. [CrossRef] [PubMed]
19. Salvi, G.E.; Fürst, M.M.; Lang, N.P.; Persson, G.R. One-Year Bacterial Colonization Patterns of Staphylococcus Aureus and Other Bacteria at Implants and Adjacent Teeth. *Clin. Oral Implant Res.* **2008**, *19*, 242–248. [CrossRef] [PubMed]
20. Sanz-Martin, I.; Doolittle-Hall, J.; Teles, R.P.; Patel, M.; Belibasakis, G.N.; Hämmerle, C.H.F.; Jung, R.E.; Teles, F.R.F. Exploring the Microbiome of Healthy and Diseased Peri-Implant Sites Using Illumina Sequencing. *J. Clin. Periodontol.* **2017**, *44*, 1274–1284. [CrossRef] [PubMed]
21. Koyanagi, T.; Sakamoto, M.; Takeuchi, Y.; Maruyama, N.; Ohkuma, M.; Izumi, Y. Comprehensive Microbiological Findings in Peri-Implantitis and Periodontitis. *J. Clin. Periodontol.* **2013**, *40*, 218–226. [CrossRef]
22. Kumar, P.S.; Mason, M.R.; Brooker, M.R.; O'Brien, K. Pyrosequencing Reveals Unique Microbial Signatures Associated with Healthy and Failing Dental Implants. *J. Clin. Periodontol.* **2012**, *39*, 425–433. [CrossRef] [PubMed]
23. Apatzidou, D.; Lappin, D.F.; Hamilton, G.; Papadopoulos, C.A.; Konstantinidis, A.; Riggio, M.P. Microbiome of Peri-Implantitis Affected and Healthy Dental Sites in Patients with a History of Chronic Periodontitis. *Arch. Oral Biol.* **2017**, *83*, 145–152. [CrossRef] [PubMed]
24. Rakic, M.; Grusovin, M.G.; Canullo, L. The Microbiologic Profile Associated with Peri-Implantitis in Humans: A Systematic Review. *Int. J. Oral Maxillofac. Implant.* **2016**, *31*, 359–368. [CrossRef]
25. Dabdoub, S.M.; Tsigarida, A.A.; Kumar, P.S. Patient-Specific Analysis of Periodontal and Peri-Implant Microbiomes. *J. Dent. Res.* **2013**, *92*, 168S–175S. [CrossRef] [PubMed]
26. Heuer, W.; Kettenring, A.; Stumpp, S.N.; Eberhard, J.; Gellermann, E.; Winkel, A.; Stiesch, M. Metagenomic Analysis of the Peri-Implant and Periodontal Microflora in Patients with Clinical Signs of Gingivitis or Mucositis. *Clin. Oral Investig.* **2012**, *16*, 843–850. [CrossRef]
27. Mombelli, A.; Décaillet, F. The Characteristics of Biofilms in Peri-Implant Disease. *J. Clin. Periodontol.* **2011**, *38*, 203–213. [CrossRef] [PubMed]
28. Mombelli, A.; van Oosten, M.A.; Schurch, E.; Land, N.P. The Microbiota Associated with Successful or Failing Osseointegrated Titanium Implants. *Oral Microbiol. Immunol.* **1987**, *2*, 145–151. [CrossRef] [PubMed]
29. Qin, D. Next-Generation Sequencing and Its Clinical Application. *Cancer Biol. Med.* **2019**, *16*, 4–10. [CrossRef]
30. Chen, P.; Sun, W.; He, Y. Comparison of the Next-Generation Sequencing (NGS) Technology with Culture Methods in the Diagnosis of Bacterial and Fungal Infections. *J. Thorac. Dis.* **2020**, *12*, 4924–4929. [CrossRef]

31. Torchia, M.T.; Austin, D.C.; Kunkel, S.T.; Dwyer, K.W.; Moschetti, W.E. Next-Generation Sequencing vs Culture-Based Methods for Diagnosing Periprosthetic Joint Infection After Total Knee Arthroplasty: A Cost-Effectiveness Analysis. *J. Arthroplast.* **2019**, *34*, 1333–1341. [CrossRef]
32. Kim, H.-J.; Ahn, D.-H.; Yu, Y.; Han, H.; Kim, S.Y.; Joo, J.-Y.; Chung, J.; Na, H.S.; Lee, J.-Y. Microbial Profiling of Peri-Implantitis Compared to the Periodontal Microbiota in Health and Disease Using 16S rRNA Sequencing. *J. Periodontal Implant Sci.* **2023**, *53*, 69–84. [CrossRef]
33. Song, L.; Jiang, J.; Li, J.; Zhou, C.; Chen, Y.; Lu, H.; He, F. The Characteristics of Microbiome and Cytokines in Healthy Implants and Peri-Implantitis of the Same Individuals. *J. Clin. Med.* **2022**, *11*, 5817. [CrossRef]
34. Pallos, D.; Sousa, V.; Feres, M.; Retamal-Valdes, B.; Chen, T.; Curtis, M.; Boaventura, R.M.; Tanaka, M.H.; Salomão, G.V.D.S.; Zanella, L.; et al. Salivary Microbial Dysbiosis Is Associated With Peri-Implantitis: A Case-Control Study in a Brazilian Population. *Front. Cell. Infect. Microbiol.* **2022**, *11*, 696432. [CrossRef]
35. Barbagallo, G.; Santagati, M.; Guni, A.; Torrisi, P.; Spitale, A.; Stefani, S.; Ferlito, S.; Nibali, L. Microbiome Differences in Periodontal, Peri-Implant, and Healthy Sites: A Cross-Sectional Pilot Study. *Clin. Oral Investig.* **2022**, *26*, 2771–2781. [CrossRef] [PubMed]
36. Shi, Y.; Tong, Z.; Zhang, Y.; Si, M.; He, F. Microbial Profiles of Peri-implant Mucositis and Peri-implantitis: Submucosal Microbial Dysbiosis Correlates with Disease Severity. *Clin. Oral Implant. Res.* **2022**, *33*, 172–183. [CrossRef]
37. Polymeri, A.; Horst, J.; Buijs, M.J.; Zaura, E.; Wismeijer, D.; Crielaard, W.; Loos, B.G.; Laine, M.L.; Brandt, B.W. Submucosal Microbiome of Peri-implant Sites: A Cross-sectional Study. *J. Clin. Periodontol.* **2021**, *48*, 1228–1239. [CrossRef] [PubMed]
38. Korsch, M.; Marten, S.-M.; Stoll, D.; Prechtl, C.; Dötsch, A. Microbiological Findings in Early and Late Implant Loss: An Observational Clinical Case-Controlled Study. *BMC Oral Health* **2021**, *21*, 112. [CrossRef]
39. Komatsu, K.; Shiba, T.; Takeuchi, Y.; Watanabe, T.; Koyanagi, T.; Nemoto, T.; Shimogishi, M.; Shibasaki, M.; Katagiri, S.; Kasugai, S.; et al. Discriminating Microbial Community Structure Between Peri-Implantitis and Periodontitis With Integrated Metagenomic, Metatranscriptomic, and Network Analysis. *Front. Cell. Infect. Microbiol.* **2020**, *10*, 596490. [CrossRef] [PubMed]
40. Ghensi, P.; Manghi, P.; Zolfo, M.; Armanini, F.; Pasolli, E.; Bolzan, M.; Bertelle, A.; Dell'Acqua, F.; Dellasega, E.; Waldner, R.; et al. Strong Oral Plaque Microbiome Signatures for Dental Implant Diseases Identified by Strain-Resolution Metagenomics. *Npj Biofilms Microbiomes* **2020**, *6*, 47. [CrossRef]
41. Aleksandrowicz, P.; Brzezińska-Błaszczyk, E.; Dudko, A.; Agier, J. Archaea Occurrence in the Subgingival Biofilm in Patients with Peri-Implantitis and Periodontitis. *Int. J. Periodontics Restor. Dent.* **2020**, *40*, 677–683. [CrossRef] [PubMed]
42. Yu, X.; Chan, Y.; Zhuang, L.; Lai, H.; Lang, N.P.; Keung Leung, W.; Watt, R.M. Intra-oral Single-site Comparisons of Periodontal and Peri-implant Microbiota in Health and Disease. *Clin. Oral Implant. Res.* **2019**, *30*, 760–776. [CrossRef] [PubMed]
43. Kröger, A.; Hülsmann, C.; Fickl, S.; Spinell, T.; Hüttig, F.; Kaufmann, F.; Heimbach, A.; Hoffmann, P.; Enkling, N.; Renvert, S.; et al. The Severity of Human Peri-implantitis Lesions Correlates with the Level of Submucosal Microbial Dysbiosis. *J. Clin. Periodontol.* **2018**, *45*, 1498–1509. [CrossRef]
44. Gao, X.; Zhou, J.; Sun, X.; Li, X.; Zhou, Y. Diversity Analysis of Subgingival Microbial Bacteria in Peri-Implantitis in Uygur Population. *Medicine* **2018**, *97*, e9774. [CrossRef] [PubMed]
45. Daubert, D.; Pozhitkov, A.; McLean, J.; Kotsakis, G. Titanium as a Modifier of the Peri-Implant Microbiome Structure. *Clin. Implant. Dent. Relat. Res.* **2018**, *20*, 945–953. [CrossRef] [PubMed]
46. Al-Ahmad, A.; Muzafferiy, F.; Anderson, A.C.; Wölber, J.P.; Ratka-Krüger, P.; Fretwurst, T.; Nelson, K.; Vach, K.; Hellwig, E. Shift of Microbial Composition of Peri-Implantitis-Associated Oral Biofilm as Revealed by 16S rRNA Gene Cloning. *J. Med. Microbiol.* **2018**, *67*, 332–340. [CrossRef] [PubMed]
47. Sousa, V.; Nibali, L.; Spratt, D.; Dopico, J.; Mardas, N.; Petrie, A.; Donos, N. Peri-Implant and Periodontal Microbiome Diversity in Aggressive Periodontitis Patients: A Pilot Study. *Clin. Oral Implant. Res.* **2017**, *28*, 558–570. [CrossRef] [PubMed]
48. Yu, X.-L.; Chan, Y.; Zhuang, L.-F.; Lai, H.-C.; Lang, N.P.; Lacap-Bugler, D.C.; Leung, W.K.; Watt, R.M. Distributions of Synergistetes in Clinically-Healthy and Diseased Periodontal and Peri-Implant Niches. *Microb. Pathog.* **2016**, *94*, 90–103. [CrossRef]
49. Shiba, T.; Watanabe, T.; Kachi, H.; Koyanagi, T.; Maruyama, N.; Murase, K.; Takeuchi, Y.; Maruyama, F.; Izumi, Y.; Nakagawa, I. Distinct Interacting Core Taxa in Co-Occurrence Networks Enable Discrimination of Polymicrobial Oral Diseases with Similar Symptoms. *Sci. Rep.* **2016**, *6*, 30997. [CrossRef] [PubMed]
50. Tsigarida, A.A.; Dabdoub, S.M.; Nagaraja, H.N.; Kumar, P.S. The Influence of Smoking on the Peri-Implant Microbiome. *J. Dent. Res.* **2015**, *94*, 1202–1217. [CrossRef]
51. Jakobi, M.; Stumpp, S.; Stiesch, M.; Eberhard, J.; Heuer, W. The Peri-Implant and Periodontal Microbiota in Patients with and without Clinical Signs of Inflammation. *Dent. J.* **2015**, *3*, 24–42. [CrossRef] [PubMed]
52. Zheng, H.; Xu, L.; Wang, Z.; Li, L.; Zhang, J.; Zhang, Q.; Chen, T.; Lin, J.; Chen, F. Subgingival Microbiome in Patients with Healthy and Ailing Dental Implants. *Sci. Rep.* **2015**, *5*, 10948. [CrossRef] [PubMed]
53. Schaumann, S.; Staufenbiel, I.; Scherer, R.; Schilhabel, M.; Winkel, A.; Stumpp, S.N.; Eberhard, J.; Stiesch, M. Pyrosequencing of Supra- and Subgingival Biofilms from Inflamed Peri-Implant and Periodontal Sites. *BMC Oral Health* **2014**, *14*, 157. [CrossRef]
54. Maruyama, N.; Maruyama, F.; Takeuchi, Y.; Aikawa, C.; Izumi, Y.; Nakagawa, I. Intraindividual Variation in Core Microbiota in Peri-Implantitis and Periodontitis. *Sci. Rep.* **2014**, *4*, 6602. [CrossRef] [PubMed]

55. Tamura, N.; Ochi, M.; Miyakawa, H.; Nakazawa, F. Analysis of Bacterial Flora Associated with Peri-Implantitis Using Obligate Anaerobic Culture Technique and 16S rDNA Gene Sequence. *Int. J. Oral Maxillofac. Implant.* **2013**, *28*, 1521–1529. [CrossRef] [PubMed]
56. da Silva, E.S.C.; Feres, M.; Figueiredo, L.C.; Shibli, J.A.; Ramiro, F.S.; Faveri, M. Microbiological Diversity of Peri-Implantitis Biofilm by Sanger Sequencing. *Clin. Oral Implant. Res.* **2014**, *25*, 1192–1199. [CrossRef] [PubMed]
57. Koyanagi, T.; Sakamoto, M.; Takeuchi, Y.; Ohkuma, M.; Izumi, Y. Analysis of Microbiota Associated with Peri-Implantitis Using 16S rRNA Gene Clone Library. *J. Oral Microbiol.* **2010**, *2*, 5104. [CrossRef]
58. Faveri, M.; Gonçalves, L.F.H.; Feres, M.; Figueiredo, L.C.; Gouveia, L.A.; Shibli, J.A.; Mayer, M.P.A. Prevalence and Microbiological Diversity of Archaea in Peri-Implantitis Subjects by 16S Ribosomal RNA Clonal Analysis: Archaea in Peri-Implantitis Subjects. *J. Periodontal Res.* **2011**, *46*, 338–344. [CrossRef]
59. Valiente-Mullor, C.; Beamud, B.; Ansari, I.; Francés-Cuesta, C.; García-González, N.; Mejía, L.; Ruiz-Hueso, P.; González-Candelas, F. One Is Not Enough: On the Effects of Reference Genome for the Mapping and Subsequent Analyses of Short-Reads. *PLoS Comput. Biol.* **2021**, *17*, e1008678. [CrossRef]
60. Nie, J.; Zhang, Q.; Zheng, H.; Xu, L.; Wang, X.; Chen, F. Pyrosequencing of the Subgingival Microbiome in Peri-implantitis after Non-surgical Mechanical Debridement Therapy. *J. Periodontal Res.* **2020**, *55*, 238–246. [CrossRef] [PubMed]
61. Goodrich, J.K.; Di Rienzi, S.C.; Poole, A.C.; Koren, O.; Walters, W.A.; Caporaso, J.G.; Knight, R.; Ley, R.E. Conducting a Microbiome Study. *Cell* **2014**, *158*, 250–262. [CrossRef] [PubMed]
62. Li, J.; Quinque, D.; Horz, H.-P.; Li, M.; Rzhetskaya, M.; Raff, J.A.; Hayes, M.G.; Stoneking, M. Comparative Analysis of the Human Saliva Microbiome from Different Climate Zones: Alaska, Germany, and Africa. *BMC Microbiol.* **2014**, *14*, 316. [CrossRef]
63. Nasidze, I.; Li, J.; Schroeder, R.; Creasey, J.L.; Li, M.; Stoneking, M. High Diversity of the Saliva Microbiome in Batwa Pygmies. *PLoS ONE* **2011**, *6*, e23352. [CrossRef] [PubMed]
64. Mason, M.R.; Nagaraja, H.N.; Camerlengo, T.; Joshi, V.; Kumar, P.S. Deep Sequencing Identifies Ethnicity-Specific Bacterial Signatures in the Oral Microbiome. *PLoS ONE* **2013**, *8*, e77287. [CrossRef]
65. Wade, W.G. The Oral Microbiome in Health and Disease. *Pharmacol. Res.* **2013**, *69*, 137–143. [CrossRef]
66. Cruz, G.D.; Chen, Y.; Salazar, C.R.; Le Geros, R.Z. The Association of Immigration and Acculturation Attributes with Oral Health among Immigrants in New York City. *Am. J. Public Health* **2009**, *99*, S474–S480. [CrossRef]
67. Menon, R.K.; Gopinath, D. Eliminating Bias and Accelerating the Clinical Translation of Oral Microbiome Research in Oral Oncology. *Oral Oncol.* **2018**, *79*, 84–85. [CrossRef] [PubMed]
68. Berglundh, T.; Armitage, G.; Araujo, M.G.; Avila-Ortiz, G.; Blanco, J.; Camargo, P.M.; Chen, S.; Cochran, D.; Derks, J.; Figuero, E.; et al. Peri-Implant Diseases and Conditions: Consensus Report of Workgroup 4 of the 2017 World Workshop on the Classification of Periodontal and Peri-Implant Diseases and Conditions. *J. Clin. Periodontol.* **2018**, *45*, S286–S291. [CrossRef] [PubMed]
69. Khammissa, R.A.G.; Feller, L.; Meyerov, R.; Lemmer, J. Peri-Implant Mucositis and Peri-Implantitis: Clinical and Histopathological Characteristics and Treatment. *SADJ* **2012**, *67*, 122, 124–126. [PubMed]
70. Salvi, G.E.; Cosgarea, R.; Sculean, A. Prevalence and Mechanisms of Peri-Implant Diseases. *J. Dent. Res.* **2017**, *96*, 31–37. [CrossRef] [PubMed]
71. Sahrmann, P.; Gilli, F.; Wiedemeier, D.B.; Attin, T.; Schmidlin, P.R.; Karygianni, L. The Microbiome of Peri-Implantitis: A Systematic Review and Meta-Analysis. *Microorganisms* **2020**, *8*, 661. [CrossRef] [PubMed]
72. Carvalho, É.B.S.; Romandini, M.; Sadilina, S.; Sant'Ana, A.C.P.; Sanz, M. Microbiota Associated with Peri-Implantitis-A Systematic Review with Meta-Analyses. *Clin. Oral Implant. Res.* **2023**, 1–12. [CrossRef] [PubMed]
73. Jeon, Y.D.; Lim, Y.S.; Lee, S.; Kim, K.M.; Ryu, C.-M.; Jung, I.Y.; Ahn, M.-Y.; Ann, H.W.; Ahn, J.Y.; Ku, N.S.; et al. A Comparison Between Next-Generation Sequencing and Bacterial Culture for the Detection of Bacteria in Clinical Specimen. *Open Forum Infect. Dis.* **2015**, *2*, 1108. [CrossRef]
74. Belibasakis, G.N.; Manoil, D. Microbial Community-Driven Etiopathogenesis of Peri-Implantitis. *J. Dent. Res.* **2021**, *100*, 21–28. [CrossRef] [PubMed]
75. Hashimoto, Y.; Okada, S.; Yasuda, K.; Kawagoe, M.; Kajiya, M.; Tsuga, K. Microbial Differences between Active and Remission Peri-Implantitis. *Sci. Rep.* **2022**, *12*, 5284. [CrossRef] [PubMed]
76. Hutton, B.; Salanti, G.; Caldwell, D.M.; Chaimani, A.; Schmid, C.H.; Cameron, C.; Ioannidis, J.P.A.; Straus, S.; Thorlund, K.; Jansen, J.P.; et al. The PRISMA Extension Statement for Reporting of Systematic Reviews Incorporating Network Meta-Analyses of Health Care Interventions: Checklist and Explanations. *Ann. Intern. Med.* **2015**, *162*, 777–784. [CrossRef] [PubMed]
77. Page, M.J.; McKenzie, J.E.; Bossuyt, P.M.; Boutron, I.; Hoffmann, T.C.; Mulrow, C.D.; Shamseer, L.; Tetzlaff, J.M.; Akl, E.A.; Brennan, S.E.; et al. The PRISMA 2020 Statement: An Updated Guideline for Reporting Systematic Reviews. *Syst. Rev.* **2021**, *10*, 89. [CrossRef]
78. ROBINS-E Development Group. Risk Of Bias In Non-randomized Studies-of Exposure (ROBINS-E). *Launch Version*, 20 June 2023. Available online: https://www.riskofbias.info/welcome/robins-e-tool (accessed on 14 August 2023).

MDPI AG
Grosspeteranlage 5
4052 Basel
Switzerland
Tel.: +41 61 683 77 34

Antibiotics Editorial Office
E-mail: antibiotics@mdpi.com
www.mdpi.com/journal/antibiotics

www.ingramcontent.com/pod-product-compliance
Lightning Source LLC
Chambersburg PA
CBHW062057220926
51850CB00010B/2795
9783725859856